SAS® Software Solutions:
Basic Data Processing

Thomas Miron

The correct bibliographic citation for this manual is as follows: Miron, Thomas, *SAS® Software Solutions: Basic Data Processing*, Cary, NC: SAS Institute Inc., 1993.

SAS® Software Solutions: Basic Data Processing

Copyright © 1993 by SAS Institute Inc., Cary, NC, USA.

ISBN 1-55544-536-5

All rights reserved. Printed in the United States of America. No part of this publication may be reproduced, stored in a retrieval system, or transmitted, in any form or by any means, electronic, mechanical, photocopying, or otherwise, without the prior written permission of the publisher, SAS Institute Inc.

The SAS® System is an integrated system of software providing complete control over data access, management, analysis, and presentation. Base SAS software is the foundation of the SAS System. Products within the SAS System include SAS/ACCESS®, SAS/AF®, SAS/ASSIST®, SAS/CALC®, SAS/CONNECT®, SAS/CPE®, SAS/DMI®, SAS/EIS®, SAS/ENGLISH®, SAS/ETS®, SAS/FSP®, SAS/GRAPH®, SAS/IML®, SAS/IMS-DL/I®, SAS/INSIGHT®, SAS/LAB®, SAS/OR®, SAS/PH-Clinical®, SAS/QC®, SAS/REPLAY-CICS®, SAS/SHARE®, SAS/STAT®, SAS/TOOLKIT®, SAS DB2™, SAS/LOOKUP™, SAS/NVISION™, SAS/SQL-DS™, and SAS/TUTOR™ software. Other SAS Institute products are SYSTEM 2000®, Data Management Software, with basic SYSTEM 2000, CREATE™, Multi-User™, QueX™, Screen Writer™, and CICS interface software; NeoVisuals® software; JMP®, JMP IN®, JMP Serve®, and JMP *Design*™ software; SAS RTERM® software; and the SAS/C® Compiler and the SAS/CX® Compiler. MultiVendor Architecture™ and MVA™ are trademarks of SAS Institute Inc. SAS Video Productions℠ and the SVP logo are service marks of SAS Institute Inc. SAS Institute also offers SAS Consulting®, Ambassador Select℠ and On-Site Ambassador℠ services. *Authorline*®, *SAS Communications*®, *SAS Training*®, *SAS Views*®, the SASware Ballot®, and *Observations*™ are published by SAS Institute Inc. All trademarks above are registered trademarks or trademarks of SAS Institute Inc. in the USA and other countries. ® indicates USA registration.

The Institute is a private company devoted to the support and further development of its software and related services.

OS/2® is a registered trademark or trademark of International Business Machines Corporation.

Other brand and product names are registered trademarks or trademarks of their respective companies.

SAS Institute does not assume responsibility for the accuracy of any material presented in this book.

Acknowledgments

This book is the result of the efforts of many people. Thanks to Zoe for abiding patience. Special thanks to Acquisitions Editor David Baggett and Copy Editor Jeff Lopes for their advice, encouragement, and tireless effort. And thanks to the following people from SAS Institute Inc. for making it all happen: Deborah Blank, Mary Cole, Betsy Corning, Anne Corrigan, Brenda Kalt, Liz Malcom, Ginny Matsey, Nancy Mitchell, Sally Painter, Blanche Phillips, Julie Platt, Randy Poindexter, Wanda Verreault, Holly Whittle, and Dea Zullo.

Contents

Introduction 3

Part 1 - Getting Started

Chapter 1 - Entering Your Data 13

Chapter 2 - Reading from a File 21

Chapter 3 - Interactive Editing 31

Chapter 4 - Sorting and Grouping Data 43

Part 2 - Working with Your Data

Chapter 5 - Counting Occurrences 55

Chapter 6 - Finding Sums 63

Chapter 7 - Finding Averages 73

Chapter 8 - Grouping Data: Days into Months 81

Chapter 9 - Grouping Data: Create Your Own Groups 91

Chapter 10 - Combining SAS Data Sets 101

Chapter 11 - Selecting and Changing Data 111

Chapter 12 - Finding Unique Values 119

Chapter 13 - Finding Percentages 129

Part 3 - Presenting Your Data

Chapter 14 - Creating a Table 139

Chapter 15 - Creating a Custom Report 149

Chapter 16 - Creating Bar Charts 167

Chapter 17 - Creating Stacked Bar Charts 181

Chapter 18 - Creating Grouped Bar Charts 195

Chapter 19 - Creating Line Graphs and Plots 209

Index 221

Introduction

SAS Software Solutions: Basic Data Processing is a learn-by-example guide for new and intermediate users of SAS® software. Each chapter examines a common data processing problem. If you need to

- key in data
- read data from a file
- count occurrences
- find totals, averages, or percentages
- generate a report or table
- create a chart or graph

you will find working examples in this book. The solution to each problem is presented in a fully annotated SAS program along with tips, technical notes, alternative methods, and a guide to related SAS Institute documentation. The goal of *SAS Software Solutions* is to help you learn how the SAS System works so you can develop SAS software solutions to your own data processing problems.

About the Author

Thomas Miron has over ten years of experience with the SAS System in applications development, technical support, consulting, technical writing, and user training. *SAS Software Solutions* is a result of this experience helping users apply the power of the SAS System.

How to Use This Book

SAS Software Solutions is arranged in three parts:

Part 1 - Getting Started covers data entry and basic report writing. It shows you how to enter, sort, and print your data.

Part 2 - Working with Your Data shows you how to analyze your information with counts, averages, totals, and data groups.

Part 3 - Presenting Your Data introduces some of the many presentation options available to you with SAS software. Tables, customized reports, and graphics are covered.

Each chapter in the book deals with a specific data processing project. Within each chapter are the following sections:

 **Problem** presents the data processing task and associated data.

 **Solution** demonstrates how to use SAS software to solve the problem. The Solution section includes a complete, working SAS program and detailed program notes that explain how and why each program statement is used.

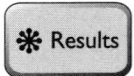 **Results** shows you the final output, printed report, or graphic produced by the SAS program.

 **Closer Look** footnotes cover program statements and related items in more detail.

 **End Notes** cover alternative methods and technical details.

 **Quick Summary** briefly outlines the problem-solving procedure.

 **More Information** lists related SAS Institute documentation.

You can use this book as a learning guide by going through the chapters in sequence. Or, when you need to solve a specific problem, skip to the examples that cover the features you need. Use the solution programs as templates and modify them to meet your needs. When you need to jog your memory on the basic steps involved, check the Quick Summary section near the end of each chapter.

Examples Used in This Book

To use the examples in this book, you must be able to run the SAS System at your site. All examples used in this book are working SAS programs written and tested by the author. The programs were executed with SAS software Release 6.06 running under the OS/2® operating system, and Release 6.08 running under the Windows operating system. Programs were created and submitted in the SAS Display Manager System.

All example programs should run on your SAS System installation without modification except for FILENAME and LIBNAME statements that refer to specific filenames and SAS data libraries, and GOPTIONS DEVICE= statements that refer to specific graphics output devices.

The example in Chapter 3 requires SAS/FSP® software. Examples in Chapters 16 through 19 require SAS/GRAPH® software. To view graphics output you need to be working at a display with graphics capability or have access to a graphics hardcopy device.

What You Need to Know

To get the most from the examples in this book, you should be familiar with how files are named and located on your computer system. At some sites you may need permission to create and read files.

If you are new to the SAS System, an excellent complement to this book is *Introducing the SAS System, Version 6, First Edition*, available from SAS Institute. *Introducing the SAS System* gives an overview of SAS software and shows you how to write and run programs with the SAS Display Manager System.

The SAS System

The SAS System consists of several integrated products from SAS Institute Inc. It is a comprehensive information delivery system that provides a set of tools for both general purpose and highly specialized data processing tasks. No other software has the breadth of applications and runs on such a wide range of computer systems. This means there is a SAS software solution for just about any data processing problem, from data entry to three-dimensional graphical analysis.

The main component of the SAS System is called base SAS software. The DATA step is a powerful application programming language within base SAS software. Base SAS software also includes many complete applications called procedures, or PROCs. These procedures can be used to print reports, generate statistics, and manage data. You can create sophisticated data processing systems by combining DATA step programs and PROC steps.

Other SAS product modules handle graphics (SAS/GRAPH software), statistical analysis (SAS/STAT® software), forecasting and modeling (SAS/ETS® software), and interactive data entry (SAS/FSP software). Several other modules are available as well. Programs in this book use base SAS, SAS/GRAPH, and SAS/FSP software.

SAS Programs

SAS programs are made up of "steps". A program can contain one or hundreds of steps. There are two kinds of SAS steps: DATA steps, and procedure or PROC steps. Programs can also contain global statements that affect such things as titles and system options. Each step or global statement is executed in sequence.

SAS programs are made up of steps and global statements. Each is executed in sequence.

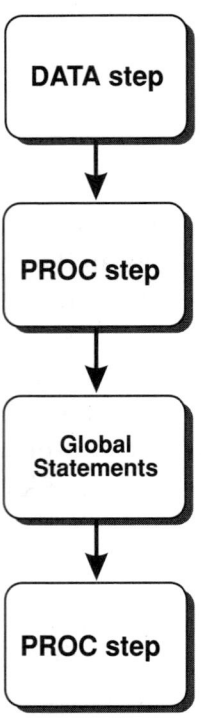

The DATA step gives you the flexibility of a programming language like COBOL, BASIC, or Pascal. You use the DATA step language to create your own customized applications. The sequence of statements within the DATA step determines the logic and overall function of the step. DATA step programs can be short and simple, or they can be used to create large and sophisticated data processing applications.

PROC steps are programs written for you by SAS Institute. Each PROC performs a specific data processing function. For example, there are procedures that sort data, perform regression analysis, create charts, count occurrences, or create full-color maps. In a single SAS program you can use any number of PROC steps together or in combination with DATA steps.

In addition to DATA steps and PROC steps, SAS programs may contain global statements. These statements are not part of any one step, but control options that affect all steps that follow them. For example, titles for output, the number of lines printed on a page, and the colors used in graphics can all be set with global statements.

A typical SAS program might begin with a DATA step followed by one or more PROC steps and global statements. The following SAS program is made up of a DATA step, a PROC MEANS step, a global TITLE statement, and a PROC PRINT step:

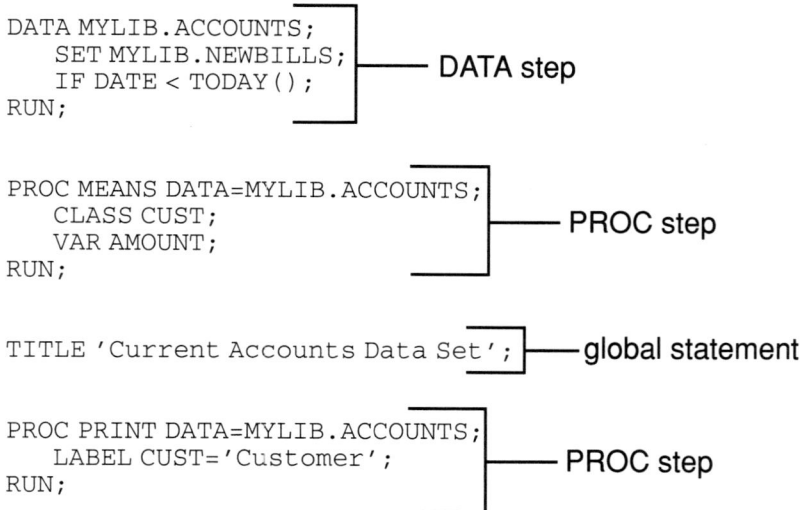

```
DATA MYLIB.ACCOUNTS;
    SET MYLIB.NEWBILLS;
    IF DATE < TODAY();
RUN;
```
— DATA step

```
PROC MEANS DATA=MYLIB.ACCOUNTS;
    CLASS CUST;
    VAR AMOUNT;
RUN;
```
— PROC step

```
TITLE 'Current Accounts Data Set';
```
— global statement

```
PROC PRINT DATA=MYLIB.ACCOUNTS;
    LABEL CUST='Customer';
RUN;
```
— PROC step

SAS Data Sets

Most SAS PROCs work only with SAS data sets. A SAS data set is a file or group of files organized so it can be processed efficiently by the SAS System. While the actual structure of a SAS data set may vary, you can think of it as a two-dimensional table consisting of rows called observations and columns called variables. An individual data item is held as the value of a variable in an observation. Most data formats can be converted to SAS data sets, including spreadsheets, database tables, and text files.

SAS data sets are organized as observations (rows) and variables (columns).

Variables

	ID	NAME	TEAM	PROJECT
1	1013	Morris	Sales	Northstar
2	1026	Fisher	Design	Vega
3	1065	Kraemer	Design	Vega
4	1091	Patik	Prototype	Northstar

Observations

The value of NAME in observation 3 is "Kraemer".

Introduction 7

Writing SAS Programs

Your SAS program is made up of one or more SAS statements. Following are some rules for writing these statements:

- Statements begin with a keyword and end with a semicolon.
- Statements can be broken over two or more lines, or two or more statements may be written on the same line.
- Uppercase and lowercase letters can be used in any combination.
- Any number of blank lines can appear between statements.
- Statements may be indented to start at any position on a line.

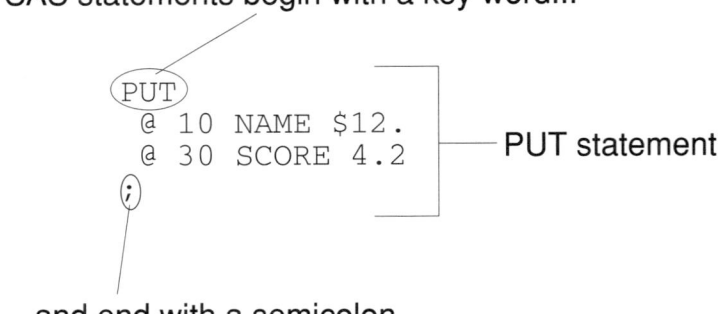

These rules leave you great flexibility to develop your own programming style. Whether or not you choose to write your programs in the style used in this book or in other SAS System documentation, the important thing is to choose a style that makes sense to you and be consistent.

The SAS Display Manager System

You can write and submit SAS statements through the SAS Display Manager System. (SAS statements can also be submitted from your computer system command line, and through a "batch job".) The display manager allows you to enter and submit program text, and view output, the SAS log, and graphics. It also provides several data management functions. See Chapter 17, "SAS Display Manager Windows," Chapter 18, "SAS Display Manager Commands," and Chapter 19, "SAS Text Editor Commands" in *SAS Language: Reference, Version 6, First Edition* for full documentation on the display manager.

SAS Display Manager System PROGRAM EDITOR window

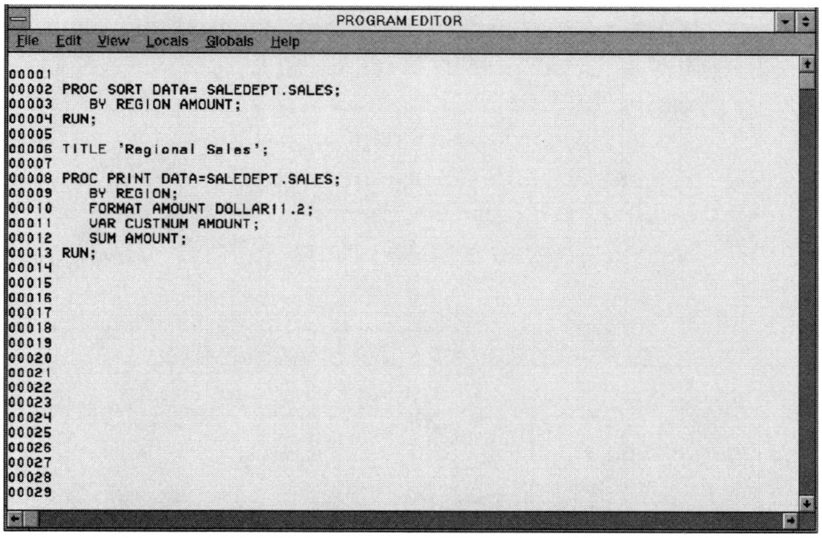

```
00001
00002 PROC SORT DATA= SALEDEPT.SALES;
00003    BY REGION AMOUNT;
00004 RUN;
00005
00006 TITLE 'Regional Sales';
00007
00008 PROC PRINT DATA=SALEDEPT.SALES;
00009    BY REGION;
00010    FORMAT AMOUNT DOLLAR11.2;
00011    VAR CUSTNUM AMOUNT;
00012    SUM AMOUNT;
00013 RUN;
```

Working Interactively

Some SAS PROCs provide interactive or "windowing" work environments. The FSEDIT procedure used in Chapter 3 is an example of an interactive procedure.

Part of the windowing interface of PROC FSEDIT

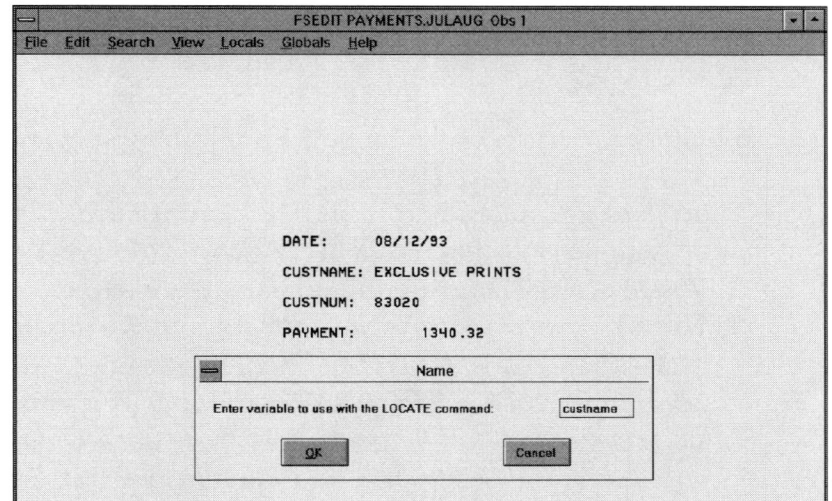

Introduction 9

The SAS Log

When you run a SAS program, information about the progress of the program, files used, and errors is recorded in the SAS log. The SAS log is an important feature of the SAS System. Not only does it help you debug your programs, but it acts as a record of SAS data set updates, files read or changed, the time and date of your run, and the amount of time each step took to execute. You can view the SAS log in the display manager LOG window. For more on the SAS log, see "SAS Log" in Chapter 5, "SAS Output," *SAS Language: Reference*.

The SAS log gives you a record of your program, including errors.

```
70    LIBNAME SALEDEPT 'C:\DATA\SAS';
NOTE: Libname SALEDEPT refers to the same physical library as PAYMENTS.
NOTE: Libref SALEDEPT was successfully assigned as follows:
      Engine:        V608
      Physical Name: C:\DATA\SAS
71    PROC SORT DATA= SALEDEPT.SALES;
72       BY REGION AMOUNT;
73    RUN;

NOTE: The data set SALEDEPT.SALES has 30 observations and 4 variables.
NOTE: The PROCEDURE SORT used 0.28 seconds.

74
75    TITLE 'Regional Sales';
76
77    PROC PRINT DATA=SALEDEPT.SALES;
78       BY REGION;
79       FOMRAT AMOUNT DOLLAR11.2;
         ------
         14
80       VAR CUSTNUM AMOUNT;
81       SUM AMOUNT;
82    RUN;
WARNING 14-169: Assuming the symbol FORMAT was misspelled as FOMRAT.

NOTE: The PROCEDURE PRINT used 0.38 seconds.
```

What Next?

There is much more to the SAS System than is covered in this book. *SAS Software Solutions: Basic Data Processing* will help you get started on your data processing projects, but you will probably need some of the other documentation referred to in the following chapters. *SAS Language: Reference* and the *SAS Procedures Guide, Version 6, First Edition* are the primary reference volumes for base SAS software. The *SAS Companion* series for your computer operating system shows you how to deal with filenames and other operating system-specific features. Each SAS product module, such as SAS/GRAPH and SAS/FSP software, has at least one associated reference volume. In addition, there are several usage guides, technical reports, and other Books by Users volumes available. All documentation referred to in this book is published by SAS Institute Inc. A complete list of available documentation is available from the Institute.

The SAS System is a powerful data processing tool with a wealth of features and options. Be prepared to take your time and experiment a little...and enjoy!

Part 1 - Getting Started

Chapter 1 - Entering Your Data

Chapter 2 - Reading from a File

Chapter 3 - Interactive Editing

Chapter 4 - Sorting and Grouping Data

Chapter 1
Entering Your Data

Learn How to...

- Enter data into your computer system
- Read character and numeric data
- Handle long character values
- Print a report with titles and column totals

Using These SAS System Features...

- SAS DATA step
- CARDS statement and instream data
- LENGTH statement
- PRINT procedure with the SUM statement

 **Problem** # Enter and Print Data from a Handwritten Form

For each completed sale, sales representatives fill out an activity form with their last name, customer number, and the amount of the sale. The forms are shown below. You need to produce a sales activity report that lists the information for each sale and includes the total of all sales.

Sales activity forms

Sales Representative	Customer Number	Amount of Sale
Thomas	30450	$3670.45
Thomas	40990	$958.89
Thomas	30450	$1730.45
Thomas	35118	$7236.03

Sales Representative	Customer Number	Amount of Sale
Wu	87764	$9761.00
Wu	74410	$756.00

Sales Representative	Customer Number	Amount of Sale
Walters	11833	$5723.12

Sales Representative	Customer Number	Amount of Sale
Christopherson	45280	$3812.78
Christopherson	45280	$1339.00
Christopherson	45280	$123.78

 # DATA Step, Instream Data, and PROC PRINT

When the data you need are on paper, your first task is to get the information into the computer system so you can work with it. One of the easiest ways to do this is with the SAS DATA step and instream data. Instream data means you manually enter data right along with your SAS program statements. When the program runs, the DATA step reads the data lines and creates a SAS data set. The data set can then be printed using the PRINT procedure, or analyzed with other SAS procedures.

Program 1 below makes use of DATA step features that allow you to handle data as numeric values or strings of characters. The PRINT procedure in Program 1 uses the SUM statement to generate the sales total. See Program 1 Notes for details.

Program 1

```
1   DATA SALES;
2
3      LENGTH NAME $ 20;
4
5      INPUT NAME $ CUSTNUM $ AMOUNT;
6
7   CARDS;
8   Thomas 30450 3670.45
9   Thomas 40990 958.89
10  Thomas 30450 1730.45
11  Thomas 35118 7236.03
12  Wu 87764 9761.00
13  Wu 74410 756.00
14  Walters 11833 5723.12
15  Christopherson 45280 3812.78
16  Christopherson 45280 1339.00
17  Christopherson 45280 123.78
18  ;
19
20  RUN;
21
22  TITLE 'Sales List';
23
24  PROC PRINT DATA=SALES;
25     SUM AMOUNT;
26  RUN;
```

Chapter 1 - Entering Your Data 15

Program 1 Notes

Line

1 The DATA statement begins the DATA step and names the SAS data set to be created, SALES.

3 The LENGTH statement says that information for the variable NAME can be up to 20 characters long. If you do not include a LENGTH statement, the SAS System assumes that character-type variables are a maximum of 8 characters long. Without the LENGTH statement, the name "Christopherson" would be truncated to "Christop".

5 The INPUT statement reads data into 3 variables: NAME, CUSTNUM, and AMOUNT. The "$" after NAME and CUSTNUM indicates that these are character variables. The SAS System assumes variables are numeric by default. AMOUNT is not followed by "$" so it is treated as a numeric variable.

7 Everything that follows this CARDS statement, up to the semicolon on line 18, is data, not SAS program statements.

8-17 These are your data. Column alignment is not critical. Be sure to type at least one space between each field.

18 The semicolon terminates the instream data. All statements that follow are again interpreted as SAS program statements.

20 The RUN statement ends and executes the DATA step.

22 The TITLE statement defines a title for printed output. The quoted text will be printed at the top of each page of output from the following PRINT procedure.

24 The PROC PRINT statement starts the PRINT procedure. The SALES data set, created in the previous step, is named as the input SAS data set.

25 The SUM statement is optional with PROC PRINT. This SUM statement prints the total of the variable AMOUNT at the end of the report. You can list one or more numeric variables in a SUM statement.

26 The RUN statement ends and executes the PROC PRINT step.

 The data set SALES, created in this example, is a **temporary SAS data set**. This means that when your SAS session ends, it will be automatically deleted. You can make a SAS data set permanent by saving it in a permanent library. For more on permanent libraries, see Chapter 6, "SAS Files," in *SAS Language: Reference, Version 6, First Edition* or Chapter 2 in this book.

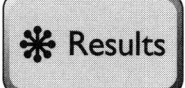 ## Sales Activity List

The resulting sales activity list is shown below. Note the title and how PROC PRINT formats the information centered on the page. Also note the result of using the SUM statement: the total of all sales is printed under the AMOUNT column.

The combination of a DATA step with instream data and PROC PRINT makes it easy to generate reports from manually entered data.

PROC PRINT output, the sales activity list

```
                        Sales List

        OBS    NAME              CUSTNUM      AMOUNT

         1     Thomas             30450       3670.45
         2     Thomas             40990        958.89
         3     Thomas             30450       1730.45
         4     Thomas             35118       7236.03
         5     Wu                 87764       9761.00
         6     Wu                 74410        756.00
         7     Walters            11833       5723.12
         8     Christopherson     45280       3812.78
         9     Christopherson     45280       1339.00
        10     Christopherson     45280        123.78
                                             ========
                                             35111.50
```

Instream Data, Input Styles

 What is instream data? In the SAS environment, manually entered data are usually referred to as *instream data*. This is because you key the information directly into the stream of SAS program statements. Your program includes both SAS statements and data.

Instream data dates back to the days of controlling computers with punched cards. Each SAS statement was punched on a single card. Data records were also punched on cards and fed into the computer right along with (instream with) the program cards. While cards are rarely used today, some of the limitations of punched cards still apply to the use of instream data. Most notable is the fact that, like the 80-column punched cards, instream data lines are limited to a maximum of 80 characters per line unless the SAS System option NOCARDIMAGE is in effect.

Instream data limitations. If you are designing an application that will require copying data from paper and be used repeatedly, you should avoid instream data as demonstrated in this chapter. The instream data method is simple and fast, but it requires that you directly edit the SAS program file. Also keep in mind that there is no interactive data validity checking. You won't know if a value was entered incorrectly until you run the program and check the output. Systems that will be used repeatedly or require data validation should use the FSEDIT procedure of SAS/FSP ® or SAS/ASSIST ® software, or another entry system that has the capacity to check data as they are entered and allows you to correct mistakes immediately.

Input styles. The style of input used in Program 1 is called *list input* because variables receive data in the order they are listed: NAME gets the first piece of information, CUSTNUM the second, and AMOUNT the third. When using list input the system assumes that a blank space separates each field. This causes problems when blanks are embedded in the data, as in the name "Mary Ann". There are several ways to overcome this limitation. One way is to use formatted input as shown in Chapter 2, "Reading from a File." Also see "Choosing the Style of Input" in Chapter 3, "Starting with Raw Data," in *SAS Language and Procedures: Usage, Version 6, First Edition*. For full documentation of INPUT statement options, refer to Chapter 9, "SAS Language Statements," *SAS Language: Reference, Version 6, First Edition*.

Entering Your Data

1. Write a SAS DATA step with an INPUT statement that names each data field. Name the output SAS data set in the DATA statement.

2. Use a CARDS statement to signal the beginning of your data.

3. Enter the data with each field separated by a blank.

4. Use a semicolon to signal the end of data.

5. Use PROC PRINT to print your report. The SUM statement in PROC PRINT prints totals for numeric variables.

DATA Step, PROC PRINT, SAS Data Libraries

character variables and values
    Chapter 7, "Working with Character Variables," *SAS Language and Procedures: Usage, Version 6, First Edition;*
    "Character Values" in Chapter 3, "Components of the SAS Language," *SAS Language: Reference, Version 6, First Edition*

DATA statement
    "DATA" in Chapter 9, "SAS Language Statements," *SAS Language: Reference*

DATA step
    "How the DATA Step Works: a Basic Introduction" in Chapter 2, "Introduction to DATA Step Processing," *SAS Language and Procedures: Usage;*
    Chapter 2, "The DATA Step," *SAS Language: Reference*

INPUT statement and list input
    "Choosing the Style of Input" in Chapter 3, "Starting with Raw Data," *SAS Language and Procedures: Usage;*
    "INPUT" in Chapter 9, "SAS Language Statements," *SAS Language: Reference*

LENGTH statement and length of character variables
    "Setting the Length of Character Variables" in Chapter 7, "Working with Character Variables," *SAS Language and Procedures: Usage;*
    "LENGTH" in Chapter 9, "SAS Language Statements," *SAS Language: Reference*

numeric values and variables
    Chapter 6, "Working with Numeric Variables," *SAS Language and Procedures: Usage;*
    "Numeric Values" in Chapter 3, "Components of the SAS Language," *SAS Language: Reference*

PROC PRINT and the SUM statement
    Chapter 27, "The Print Procedure," *SAS Procedures Guide, Version 6, Third Edition*

SAS data libraries, permanent and temporary
    Chapter 31, "SAS Data Libraries," *SAS Language and Procedures: Usage;*
    "New Concepts: SAS Data Library Model" in Chapter 6, "SAS Files," *SAS Language: Reference*

Chapter 2
Reading from a File

Learn How to...

- Read a file on your computer system
- Define input fields
- Read dates
- Print with column totals

Using These SAS System Features...

- SAS DATA step
- INFILE and INPUT statements
- SAS Informat
- FORMAT statement
- PROC PRINT

 **Problem**

Create a Patient Visit Report

At the end of the day, the office manager of a medical clinic records the date and number of patient visits for each of three clinic areas. The date, area, and number of visits are entered into a computer file. The three clinic areas are obstetrics (OB-GYN), general medicine (GENMED), and eye-ear-nose-throat (EENT). You need to read the visits file and produce a printed report listing the date, the number of visits to each area, the clinic area name, and the total number of visits to the clinic.

The file where the information is recorded is shown below. It has 15 records covering five days. The field in columns 1–8 is the date of the visit; columns 10–12 hold the number of patients; and columns 15–20 show the clinic area. The file is stored on a PC with filename `c:\clinic\visits.dat`.

The patient visit file

Record	Column 1-8 (Date)	Column 10-12 (Patients)	Column 15-20 (Area)
1	01/06/93	14	OBGYN
2	01/06/93	127	GENMED
3	01/06/93	9	EENT
4	01/07/93	3	OBGYN
5	01/07/93	83	GENMED
6	01/07/93	8	EENT
7	01/08/93	13	OBGYN
8	01/08/93	101	GENMED
9	01/08/93	12	EENT
10	01/09/93	11	OBGYN
11	01/09/93	97	GENMED
12	01/09/93	2	EENT
13	01/10/93	22	OBGYN
14	01/10/93	76	GENMED
15	01/10/93	10	EENT

22 Part 1 - Getting Started

DATA Step, INPUT Statement, PROC PRINT

Any file that is not a SAS data set or otherwise managed by the SAS System is called an *external file*. The patient visit file is an example of an external file. (See End Notes for more on external files.) Most SAS procedures, including PROC PRINT, work only with SAS data sets, not external files. To create the patient visits report with PROC PRINT, you first need to get the data into a SAS data set. There are three steps to reading an external file into a SAS data set:

1. Find the full name of the file you need to read. This name is used with a SAS FILENAME statement.

2. Determine the structure of the file. This includes the column position and length of each field, and what kind of data each field contains.

3. Write a DATA step to read the file and create the SAS data set. The DATA step includes an INFILE statement identifying the external file named on a preceding FILENAME statement, and an INPUT statement that describes each data field you want to read.

The name and layout of the patient data file, shown on the previous page, are all the information you need to write a DATA step to read the file. Once the data are in a SAS data set, you can use PROC PRINT to generate the visits report. The SAS program to create the clinic report is shown in Program 2 on the following page.

There are several operating system-dependent options that affect how the SAS System reads files. If these options are not set correctly you may encounter problems. A common problem has to do with record length. If you receive an error message that refers to the length of the record or to reading past the end of a record, check the *SAS Companion* book for your computer system for information on reading files and INFILE statement options. Also see "INFILE" in Chapter 9, "SAS Language Statements," *SAS Language: Reference, Version 6, First Edition*.

Chapter 2 - Reading from a File

Program 2

```
1    FILENAME PATDATA 'c:\clinic\visits.dat';
2
3    LIBNAME MYLIB 'c:\sasdata';
4
5    DATA MYLIB.PATIENTS;
6
7       FORMAT VISITDAY DATE7.;
8
9       INFILE PATDATA;
10
11      INPUT
12            @1    VISITDAY   MMDDYY8.
13            @10   PATIENTS   3.
14            @15   CATEGORY   $6.
15      ;
16
17   RUN;
18
19   TITLE 'Patient Count';
20
21   PROC PRINT DATA=MYLIB.PATIENTS;
22      SUM PATIENTS;
23   RUN;
```

SAS date values are stored as numeric variables representing the number of days since January 1, 1960. When you read the characters "07/04/93" with the MMDDYY8. informat, the SAS System checks to ensure the characters represent a valid date and translates to 12,238–the number of days between January 1, 1960 and July 4, 1993. Dates are stored as numbers, but there are many options for reading and displaying dates in familiar formats. For more information on handling date values see Chapter 13, "Working with Dates in the SAS System" in *SAS Language and Procedures: Usage, Version 6, First Edition*, and "SAS Informats" in Chapter 3, "Components of the SAS Language," *SAS Language: Reference, Version 6, First Edition*.

Program 2 Notes

Line **1** The FILENAME statement attaches a 1–8 character nickname to an external file. This nickname is called the *fileref.* PATDATA is the nickname attached to `c:\clinic\visits.dat`. The filename must be enclosed in quotes. Note that the FILENAME statement must come before the DATA step that contains the INFILE statement where the nickname is used.

3 The LIBNAME statement identifies the PC directory, `c:\sasdata`, as a permanent SAS library with the nickname MYLIB. MYLIB is called the *libref.* The SAS data set PATIENTS, created in the following DATA step, will be stored in this library. Data sets in permanent SAS libraries are not deleted when you end your SAS session.

The exact format of LIBNAME and FILENAME statements depends on your computer system. The formats in this example are valid for OS/2, Windows, or DOS systems.

5 The DATA statement signals the start of the DATA step and names the SAS data set that will be created, MYLIB.PATIENTS. This data set has a two-part name. The first part, MYLIB, indicates the libref of the permanent SAS library where the data set will be stored. The second part, PATIENTS, is the data set name. When referring to the data set, you must use both parts.

7 The FORMAT statement attaches the DATE7. format to the SAS date variable, VISITDAY. For example, April 20th, 1993 displayed in the DATE7. format is "20APR93". The "7" refers to the maximum number of columns available to display the data. The column length must end with a period. Without a date format, values for VISITDAY would be displayed as the number of days between Jan. 1, 1960 and the given date value.

9 The INFILE statement names the external file to use for input. The nickname (fileref) PATDATA was defined by the FILENAME statement on line 1.

The format of the file name used with the **FILENAME statement** depends on your host computer system. The file in this example works on a PC running OS/2, Windows, or PC DOS. On other systems, filenames have different formats, but the FILENAME statement syntax is similar. For details on the FILENAME statement and other operating system-specific features, see the *SAS Companion* series for your system. For example, for CMS and UNIX systems refer to the *SAS Companion for the CMS Environment, Version 6, First Edition,* and the *SAS Companion for the UNIX Environment and Derivatives, Version 6, First Edition,* respectively.

Note that you also have the option of specifying the system filename directly on the INFILE statement rather than using a FILENAME statement:

INFILE 'c:\clinic\visits.dat';

Using a nickname assigned with a FILENAME statement is usually preferred because it provides more flexibility for changes and allows for better program organization.

Chapter 2 - Reading from a File 25

Program 2 Notes

Line 11-15 This is the INPUT statement. The style of input here is called formatted input. (See Chapter 1 for an example of list input.) With *formatted input* each data field is defined by three pieces of information: the beginning column, a SAS variable name, and a SAS informat. The informat is made up of a name followed by the number of columns to read. Variable names are your choice. They must be 1–8 characters long and begin with a letter (A–Z) or underscore (_).

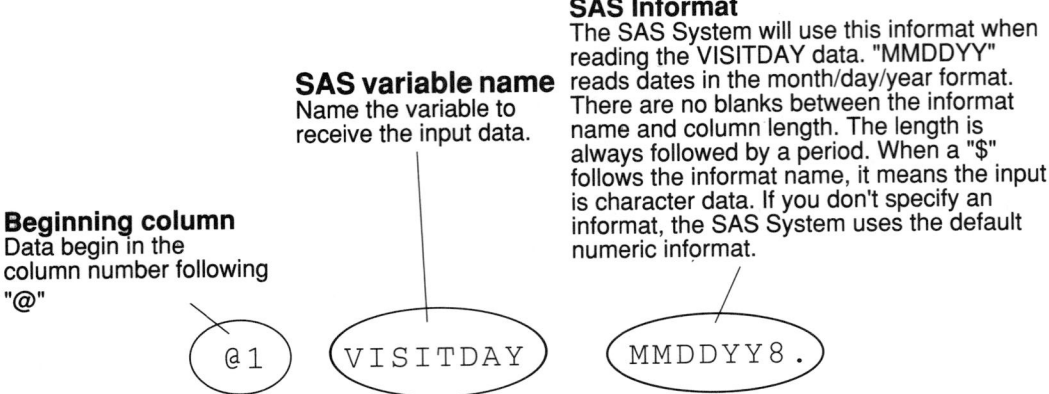

Like all SAS statements, the INPUT statement may span any number of lines. Here, the keyword INPUT, the semicolon that ends the statement, and each SAS variable are on separate lines for easier reading and editing.

17 The RUN statement ends and executes the DATA step that began on line 5.

19 The TITLE statement will apply to the following PRINT procedure. The quoted text will be printed at the top of each page of output.

21-23 The PRINT procedure step generates your report from MYLIB.PATIENTS, created in the previous DATA step. The SUM statement tells PROC PRINT to add up all the values of PATIENTS and print the total at the end of the report.

 In this example you use 3 **SAS informats**: MMDDYY for dates; $ for character values; and the default numeric informat (no informat specified) for the patient count number. Informats translate input data into standard SAS numeric or character values. With the proper informat specification, you can read a variety of data types including numbers with embedded commas, dollar signs, and percent signs. There are also informats for specialized data types such as hexadecimal numbers and datetime stamps.

Closely related to informats are formats. In this example you assigned the DATE7. format to the variable VISITDAY. Informats translate input data into SAS variable values. Formats translate SAS variable values for display or print output. There are many SAS informats and formats. For details and a list of formats and informats, see "SAS Informats" and "SAS Formats" in Chapter 3, "Components of the SAS Language," *SAS Language: Reference.*

26 Part 1 - Getting Started

Patient Visit Report

Results from Program 2 are shown below. Values for the VISITDAY variable are formatted using DATE7. as specified in the DATA step, line 7. PROC PRINT prints observation numbers in the far left column and uses variable names as column headings. The SUM statement (line 22) calculates the patient total.

The patient visit report

```
                   Patient Count

     OBS    VISITDAY    PATIENTS     CATEGORY

      1     06JAN93        14        OBGYN
      2     06JAN93       127        GENMED
      3     06JAN93         9        EENT
      4     07JAN93         3        OBGYN
      5     07JAN93        83        GENMED
      6     07JAN93         8        EENT
      7     08JAN93        13        OBGYN
      8     08JAN93       101        GENMED
      9     08JAN93        12        EENT
     10     09JAN93        11        OBGYN
     11     09JAN93        97        GENMED
     12     09JAN93         2        EENT
     13     10JAN93        22        OBGYN
     14     10JAN93        76        GENMED
     15     10JAN93        10        EENT
                       ========
                          588
```

There are several SAS system options that **control page layout**. These include:

NUMBER/NONUMBER   - prints page numbers on the page.
CENTER/NOCENTER   - centers output in the page width.
DATE/NODATE   - prints the current date on the page.

For the output shown in this book, options are set to NONUMBER, CENTER, and NODATE. For more details see "SAS System Options" in Chapter 3, "Components of the SAS Language," *SAS Language: Reference.*

Chapter 2 - Reading from a File **27**

External Files

 External files. A file that is not a SAS data set or otherwise managed by the SAS System is called an *external file*. Most external data files are organized into rows and columns, called records and fields. SAS data sets are also organized as rows and columns called observations and variables.

Most SAS procedures work only with SAS data sets, not external files. The SAS DATA step has powerful features that allow you to read external files and create SAS data sets for use with SAS procedures. The general process is shown below.

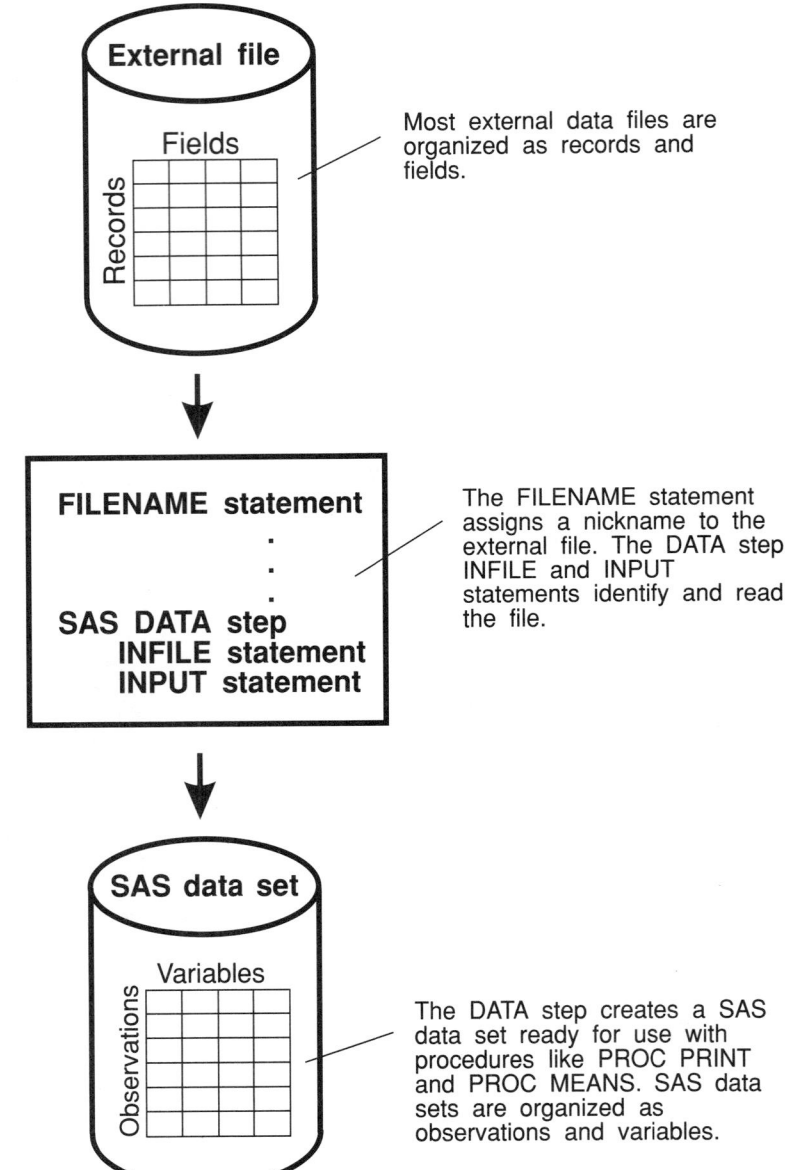

28 Part 1 - Getting Started

# Reading from a File

1. Make sure you know the name and layout of the file you want to read.

2. Use a FILENAME statement to attach a nickname (fileref) to the file you want to read.

3. Write a DATA step that names the SAS data set you want to create in the DATA statement.

4. The DATA step must contain an INFILE statement that refers to the external file, and an INPUT statement that describes the fields you want to read.

5. You can add SAS procedure steps, such as PROC PRINT, to process the resulting SAS data set.

Handling Dates and External Files, Formatting Data

dates
Chapter 13, "Working with Dates in the SAS System," *SAS Language and Procedures: Usage, Version 6, First Edition*

external files
"Special Topic: Using External Files in Your SAS Job" in Chapter 2, "Introduction to DATA Step Processing," *SAS Language and Procedures: Usage*

FILENAME statement
"FILENAME" in Chapter 9, "SAS Language Statements," *SAS Language: Reference, Version 6, First Edition*

FORMAT statement
"FORMAT" in Chapter 9, "SAS Language Statements," *SAS Language: Reference*

INFILE statement
"INFILE" in Chapter 9, "SAS Language Statements," *SAS Language: Reference*

INPUT statement
"Information You Supply to Create A SAS Data Set" in Chapter 2, "Introduction to DATA Step Processing," *SAS Language and Procedures: Usage*
"INPUT" in Chapter 9, "SAS Language Statements," *SAS Language: Reference*

SAS data sets
"The SAS Data Set: Your Key to the SAS System" in Chapter 2, "Introduction to DATA Step Processing," *SAS Language and Procedures: Usage*
"New Concepts: SAS Data Set Model" in Chapter 6, "SAS Files," *SAS Language: Reference*

SAS formats
"SAS Formats" in Chapter 3, "Components of the SAS Language," *SAS Language: Reference*

SAS informats
"SAS Informats" in Chapter 3, "Components of the SAS Language," *SAS Language: Reference*

SAS system options
"SAS System Options" in Chapter 3, "Components of the SAS Language," *SAS Language: Reference*

Chapter 3
Interactive Editing

Learn How to...

- Interactively change, add, and delete observations

Using These SAS System Features...

- FSEDIT procedure

- STRING and SEARCH commands

- ADD and DELETE commands

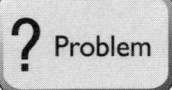

Update Customer Payment Records

The accounting department records customer payments for the months of July and August in SAS data set PAYMENTS.JULAUG, shown below. The variable DATE holds SAS date values (the number of days since Jan. 1, 1960) and is displayed in the MMDDYY8. format. CUSTNUM is the customer's ID number. CUSTNAME is the name of the customer, and PAYMENT is the dollar amount received.

You need to correct a mistaken entry for SUNTECH SOLAR in observation 5. The payment was recorded as $112.37, but it should have been $122.37. You also need to add a new payment for PAGLIA LANDSCAPE, received on August 28 for $454.21. Finally, the payment for HANSEN DEVELOPMENT, observation 3, was made in June and incorrectly entered in the JULAUG data set. You need to delete it.

SAS data set PAYMENTS.JULAUG

OBS	DATE	CUSTNUM	CUSTNAME	PAYMENT
1	08/12/93	83020	EXCLUSIVE PRINTS	1340.32
2	07/01/93	91109	SOUTHEAST SUPPLY	7645.01
3	06/28/93	84404	HANSEN DEVELOPMENT	3901.56
4	08/11/93	87811	B & B DESIGN	765.00
5	08/26/93	88028	SUNTECH SOLAR	112.37
6	08/21/93	90016	HOLMES ASSOC.	5010.96
7	07/30/93	93719	CHAVEZ PUBLISHING	2654.25
8	08/31/93	89953	THE HOLWELL GROUP	10276.45
9	08/22/93	85611	EL CENTRO SERVICES	652.00
10	08/11/93	88338	BAKER CREATIVE	1246.46
11	07/06/93	90901	SHELLY & KEATS INC.	912.10

The data set used in this example is a permanent SAS data set. The name, PAYMENTS.JULAUG, has two parts: the libref PAYMENTS, and the name JULAUG. The libref PAYMENTS refers to a permanent SAS data set library that is assumed to be available when Program 3 is run. See Chapter 2 in this book for more information on librefs, and identifying and using permanent SAS libraries.

32 Part 1- Getting Started

Solution | # FSEDIT Procedure

The FSEDIT procedure in SAS/FSP software allows you to interactively edit a SAS data set. With PROC FSEDIT you can search for, change, add, and delete observations. Program 3 shows how to start PROC FSEDIT for the payments data set, PAYMENTS.JULAUG. The screen sequence on the following pages shows you how to find and correct the SUNTECH SOLAR payment error, add a new payment for PAGLIA LANDSCAPE, and delete the HANSEN DEVELOPMENT observation.

PROC FSEDIT gives you the option of using commands or menus for most functions. This example uses menus, but the functions are also available by typing the command on a command line. To use menus, double-click on your selection with the mouse. Or, under Windows and OS/2, press the Alt key to jump to the action bar, then use the arrow keys to highlight your selection. Press the Enter key to execute the command or action. In a dialog box you can use the tab key to jump to selection buttons, then press Enter to execute. The appearance of screens and menus may vary with the type of computer system you are using.

Program 3

```
1  PROC FSEDIT DATA=PAYMENTS.JULAUG;
2  RUN;
```

Closer Look | **To use PROC FSEDIT and other SAS/FSP procedures** you must have SAS/FSP software installed on your system and be at a compatible workstation, PC, or terminal. PROC FSEDIT has many features not touched upon here. You can control screen layout and colors, use multiple screens for each observation, and create sophisticated applications with Screen Control Language (SCL). See Chapter 10, "Enhancing Data Entry Applications Using Screen Control Language" in *SAS/FSP Software: Usage and Reference, Version 6, First Edition* for more on SCL.

1. Begin at observation 1 of PAYMENTS.JULAUG.

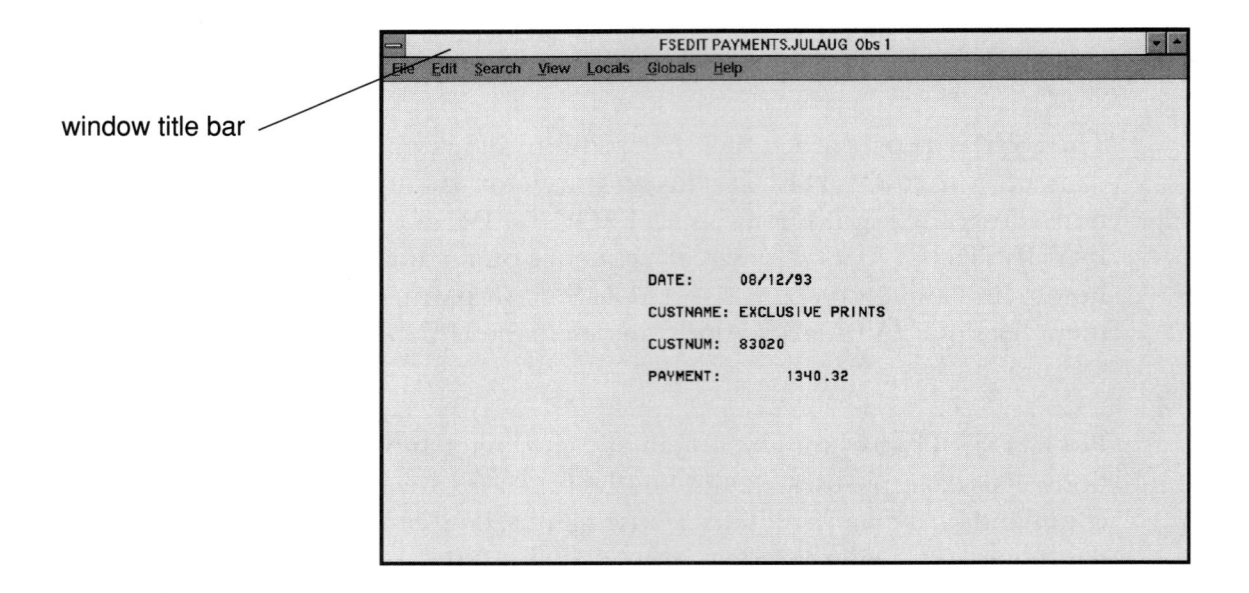

window title bar —

```
                    FSEDIT PAYMENTS.JULAUG Obs 1
  File  Edit  Search  View  Locals  Globals  Help




              DATE:     08/12/93
              CUSTNAME: EXCLUSIVE PRINTS
              CUSTNUM:  83020
              PAYMENT:     1340.32
```

When you submit Program 3 , the FSEDIT procedure displays the first observation in the PAYMENTS.JULAUG data set. The name of the data set and observation number are displayed in the window title bar.

2. Select Search and String.

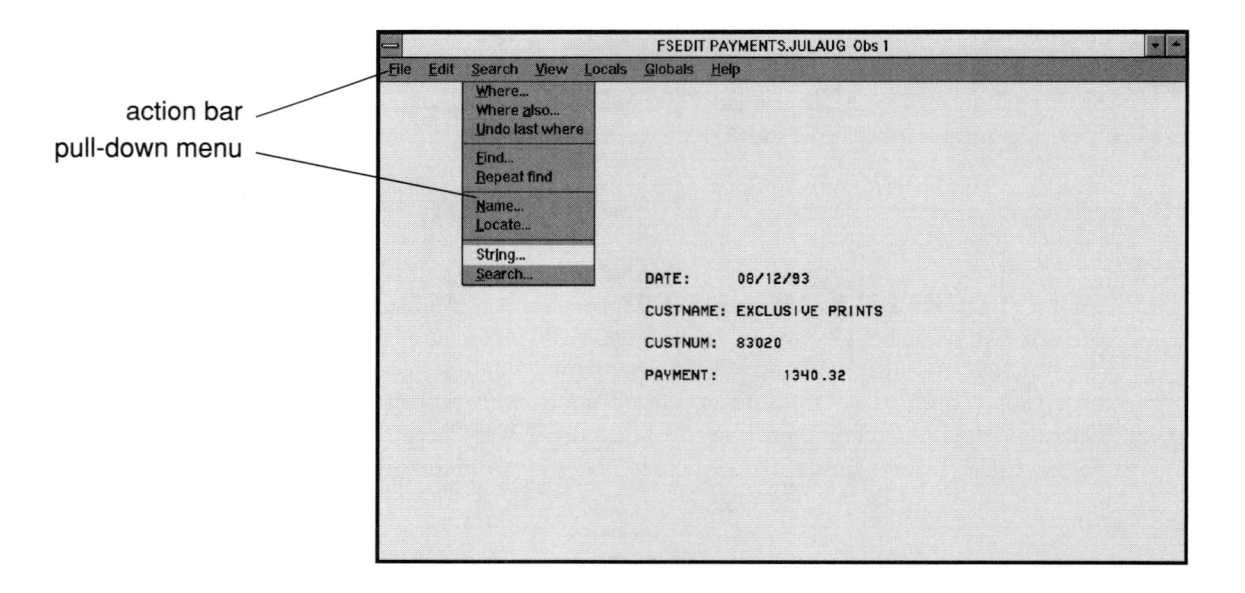

action bar —
pull-down menu —

```
                    FSEDIT PAYMENTS.JULAUG Obs 1
  File  Edit  Search  View  Locals  Globals  Help
              Where...
              Where also...
              Undo last where
              Find...
              Repeat find
              Name...
              Locate...
              String...
              Search...
                        DATE:     08/12/93
                        CUSTNAME: EXCLUSIVE PRINTS
                        CUSTNUM:  83020
                        PAYMENT:     1340.32
```

To find the SUNTECH SOLAR entry, start by selecting **Search** from the action bar. Then, from the pull-down menu, select **String...** to name the variable you want to search.

3. Enter the search variable name in the string dialog box.

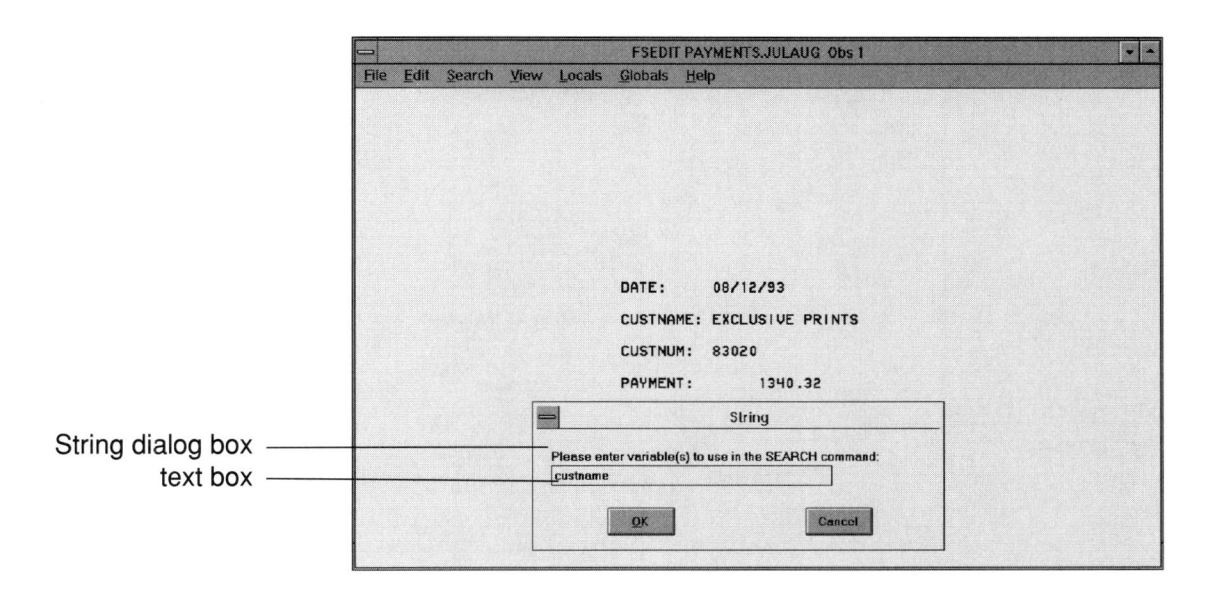

String dialog box
text box

Next, the String dialog box appears. In the text box, enter the name of the variable you want to search: CUSTNAME. The variable name is not case sensitive. Select **OK**. The Search command will now focus on the CUSTNAME variable.

4. Select the SEARCH command.

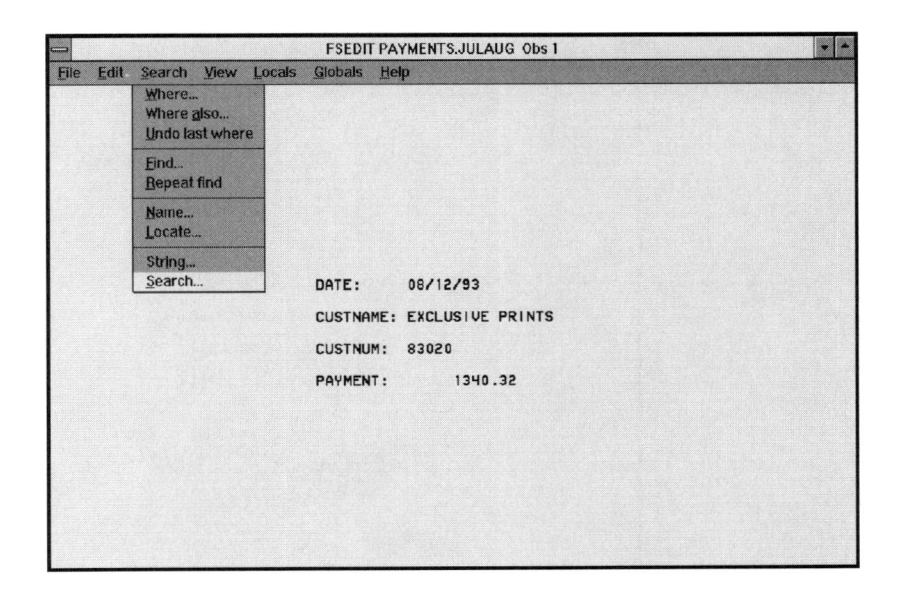

Select **Search** from the action bar. Then, select **Search...** from the pull-down menu.

Chapter 3 - Interactive Editing **35**

5. Enter the search argument in the Search dialog box.

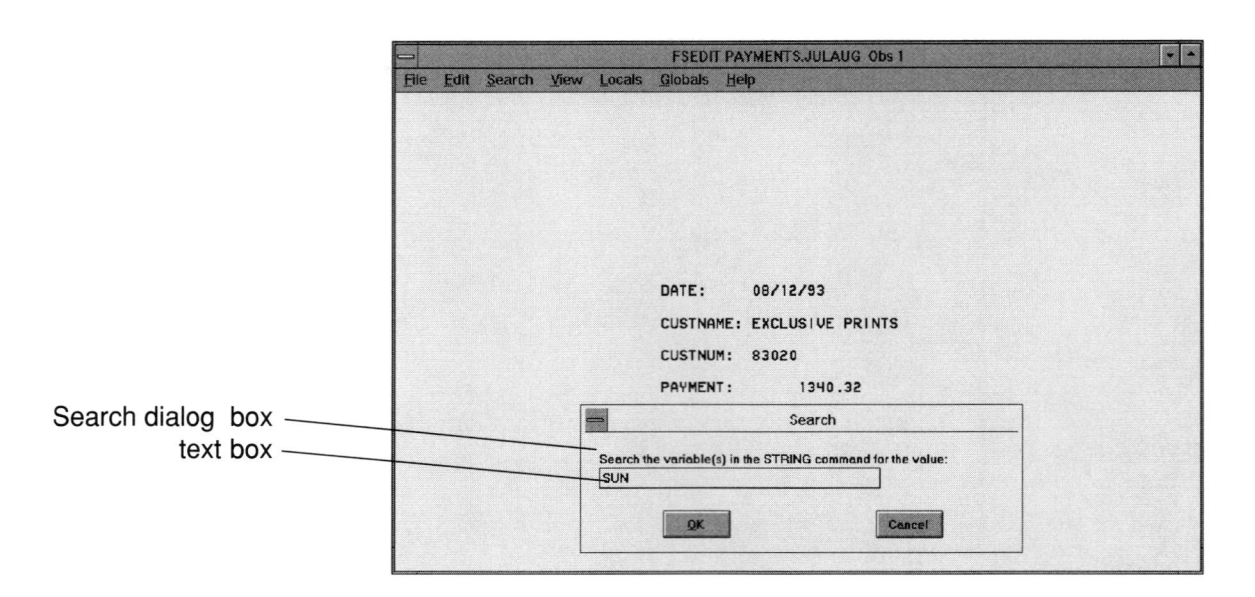

Search dialog box —
text box —

[Screenshot: FSEDIT PAYMENTS.JULAUG Obs 1 window showing DATE: 08/12/93, CUSTNAME: EXCLUSIVE PRINTS, CUSTNUM: 83020, PAYMENT: 1340.32, with a Search dialog box containing "Search the variable(s) in the STRING command for the value:" and the text "SUN" entered, with OK and Cancel buttons.]

The Search command locates character strings in SAS variables. It will locate embedded characters and partial matches. To find SUNTECH SOLAR you do not need to enter the entire name. The characters "SUN" are enough to identify this customer. Search is case sensitive and works forward from the observation following the current observation. Enter "SUN" in the text box and select OK.

6. The Search value is located and displayed.

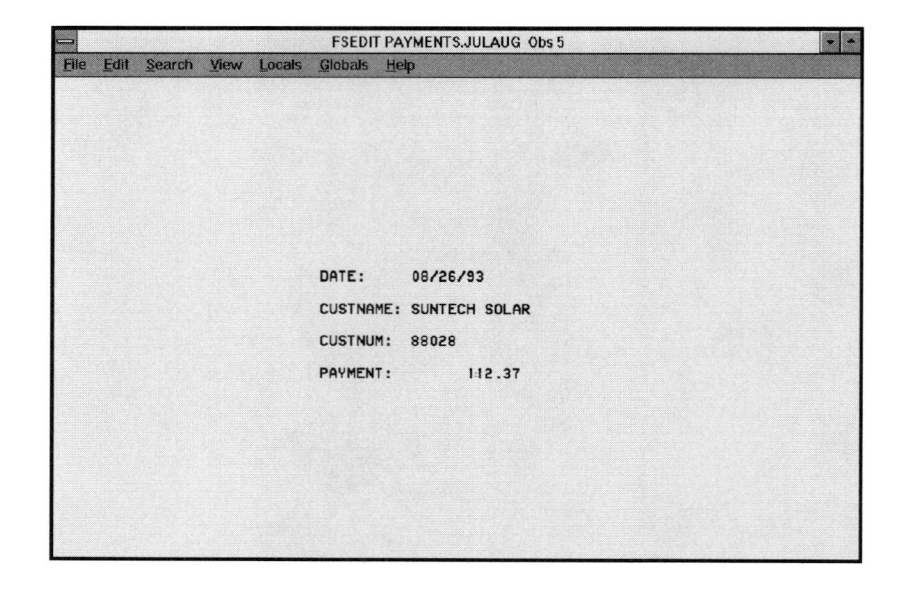

[Screenshot: FSEDIT PAYMENTS.JULAUG Obs 5 window showing DATE: 08/26/93, CUSTNAME: SUNTECH SOLAR, CUSTNUM: 88028, PAYMENT: 112.37.]

The SUNTECH SOLAR observation is located and displayed as the current observation.

36 Part 1- Getting Started

7. Correct the PAYMENT value.

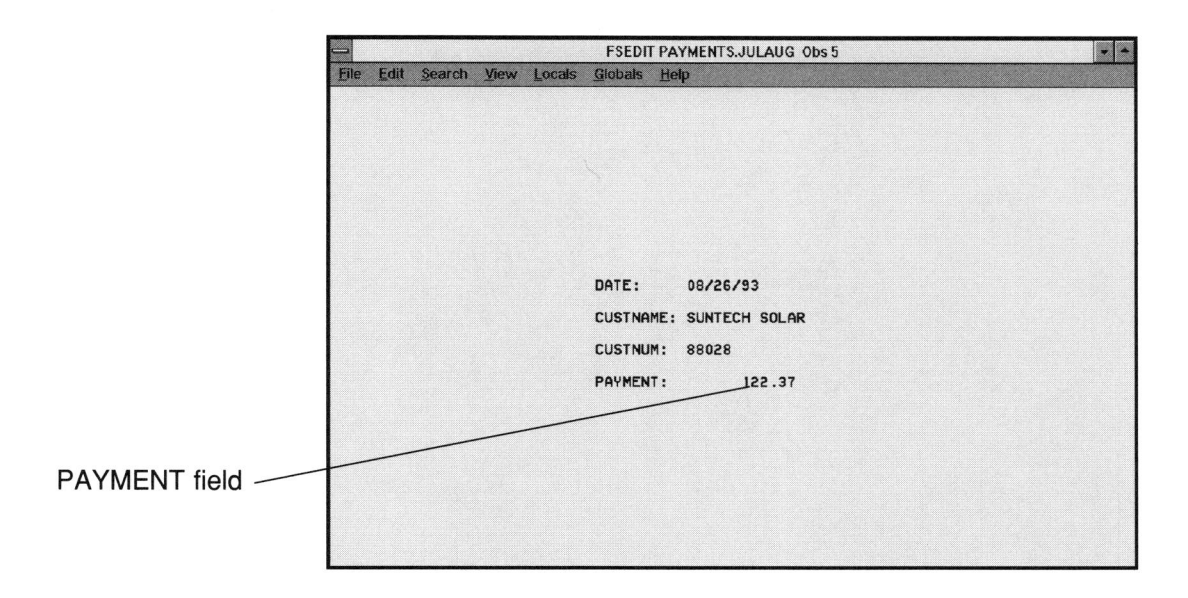

```
┌─────────────── FSEDIT PAYMENTS.JULAUG Obs 5 ───────────────┐
│ File  Edit  Search  View  Locals  Globals  Help            │
│                                                             │
│                                                             │
│                   DATE:     08/26/93                        │
│                   CUSTNAME: SUNTECH SOLAR                   │
│                   CUSTNUM:  88028                           │
│                   PAYMENT:        122.37                    │
│                                                             │
└─────────────────────────────────────────────────────────────┘
```

PAYMENT field → (points to 122.37)

To fix the SUNTECH SOLAR error, simply type the new value, 122.37, in the PAYMENT field as shown.

8. Add a new record.

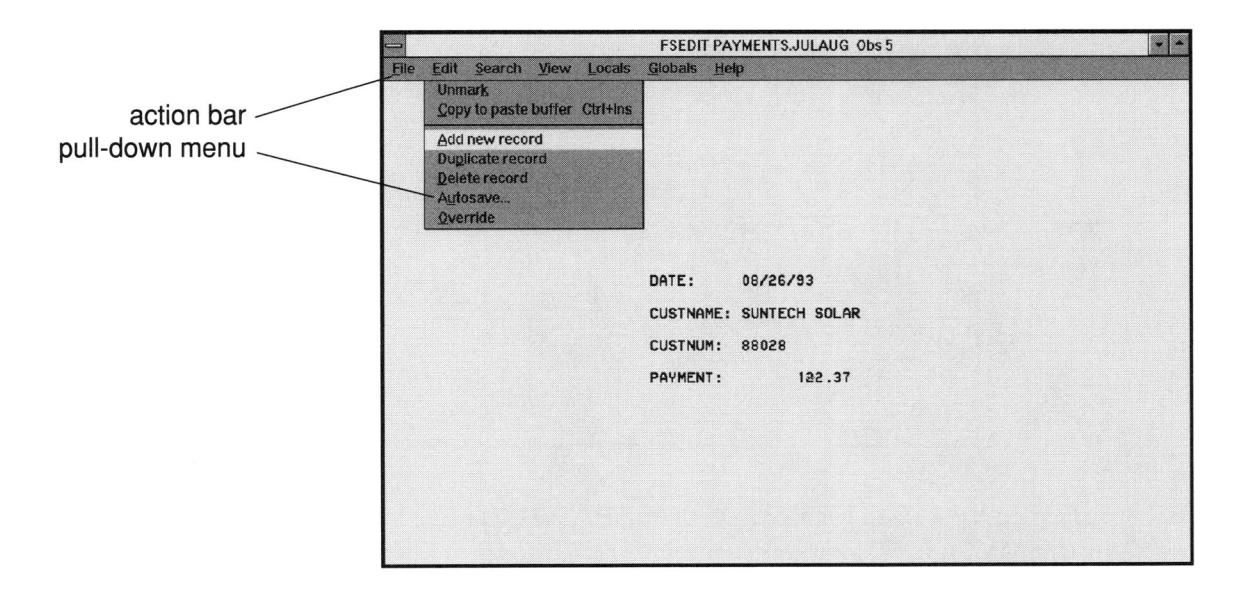

```
┌─────────────── FSEDIT PAYMENTS.JULAUG Obs 5 ───────────────┐
│ File  Edit  Search  View  Locals  Globals  Help            │
│       Unmark                                                │
│       Copy to paste buffer  Ctrl+Ins                        │
│       Add new record                                        │
│       Duplicate record                                      │
│       Delete record                                         │
│       Autosave...                                           │
│       Override                                              │
│                                                             │
│                   DATE:     08/26/93                        │
│                   CUSTNAME: SUNTECH SOLAR                   │
│                   CUSTNUM:  88028                           │
│                   PAYMENT:        122.37                    │
│                                                             │
└─────────────────────────────────────────────────────────────┘
```

action bar
pull-down menu

Next, you need to add a new payment record for PAGLIA LANDSCAPE. Select **Edit** from the action bar, then select **Add new record** from the pull-down menu.

Chapter 3 - Interactive Editing 37

9. The new observation screen appears with all fields empty.

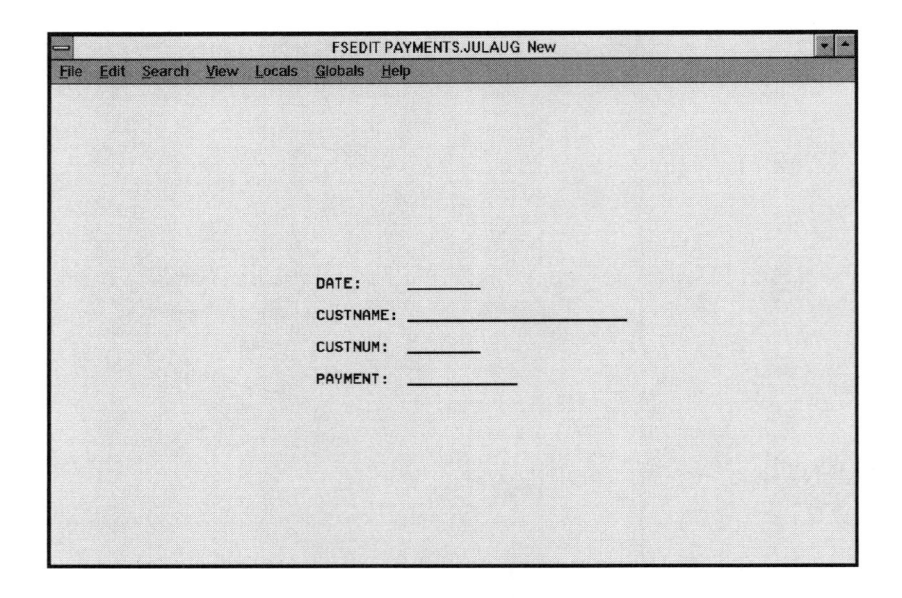

```
┌─────────────────── FSEDIT PAYMENTS.JULAUG New ───────────────────┐
│ File  Edit  Search  View  Locals  Globals  Help                  │
│                                                                  │
│                                                                  │
│                                                                  │
│                                                                  │
│                    DATE:     _____                            │
│                    CUSTNAME: _____                │
│                    CUSTNUM:  _____                             │
│                    PAYMENT:  _____                          │
│                                                                  │
│                                                                  │
│                                                                  │
└──────────────────────────────────────────────────────────────────┘
```

10. Enter the new data.

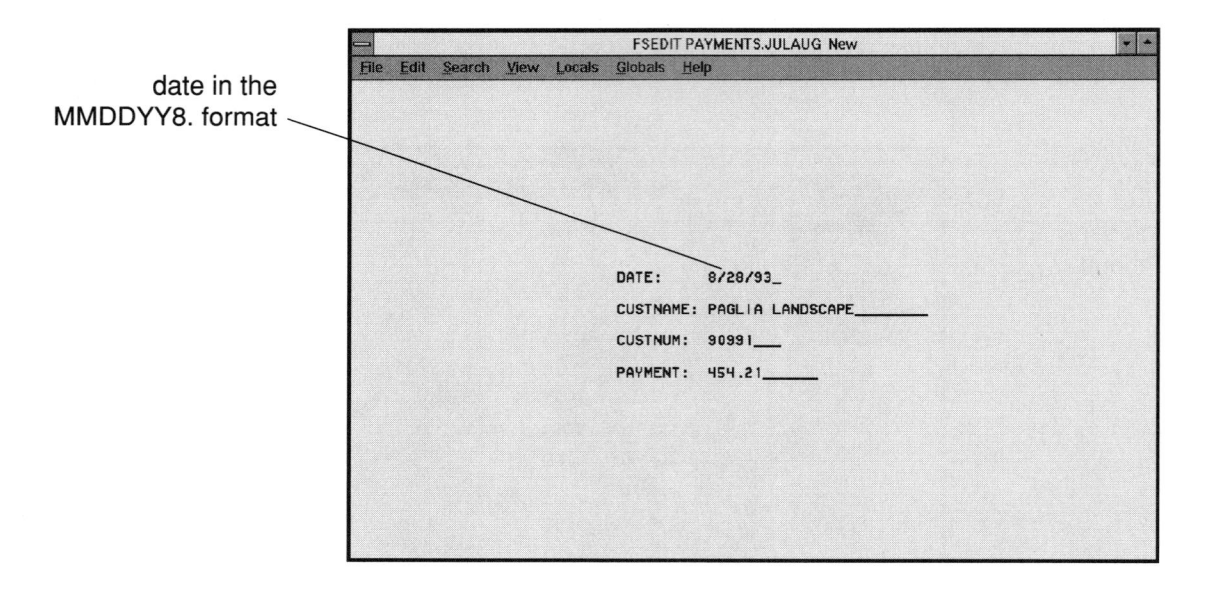

date in the MMDDYY8. format →

```
┌─────────────────── FSEDIT PAYMENTS.JULAUG New ───────────────────┐
│ File  Edit  Search  View  Locals  Globals  Help                  │
│                                                                  │
│                                                                  │
│                    DATE:     8/28/93_                            │
│                    CUSTNAME: PAGLIA LANDSCAPE_____            │
│                    CUSTNUM:  90991___                            │
│                    PAYMENT:  454.21_____                        │
│                                                                  │
└──────────────────────────────────────────────────────────────────┘
```

Enter the new data for PAGLIA LANDSCAPE. The DATE value must be entered in the MMDDYY8. format as shown. See End Notes for more on entering dates.

11. Delete an observation.

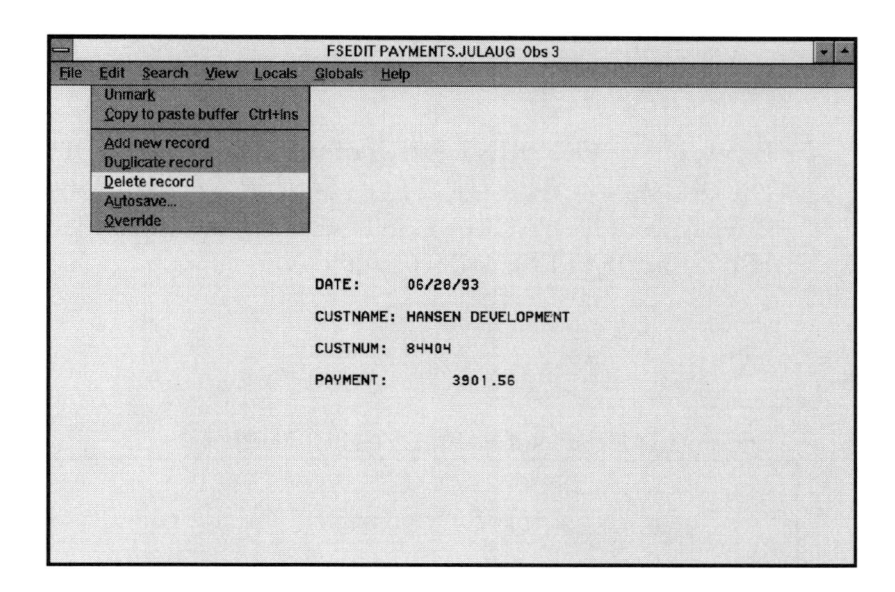

[Screenshot: FSEDIT PAYMENTS.JULAUG Obs 3 with Edit menu showing Unmark, Copy to paste buffer, Add new record, Duplicate record, Delete record, Autosave, Override. DATE: 06/28/93, CUSTNAME: HANSEN DEVELOPMENT, CUSTNUM: 84404, PAYMENT: 3901.56]

Finally, you need to delete the observation for HANSEN DEVELOPMENT. Locate the HANSEN DEVELOPMENT observation using the Search command. (See screens 2 through 5.) Or, you can scroll through observations using the Backward and Forward keys. Select **Keys** from the **Help** pull-down menu to get a list of key assignments. Next, select **Edit** from the action bar and **Delete record** from the pull-down menu.

12. Exit the FSEDIT session.

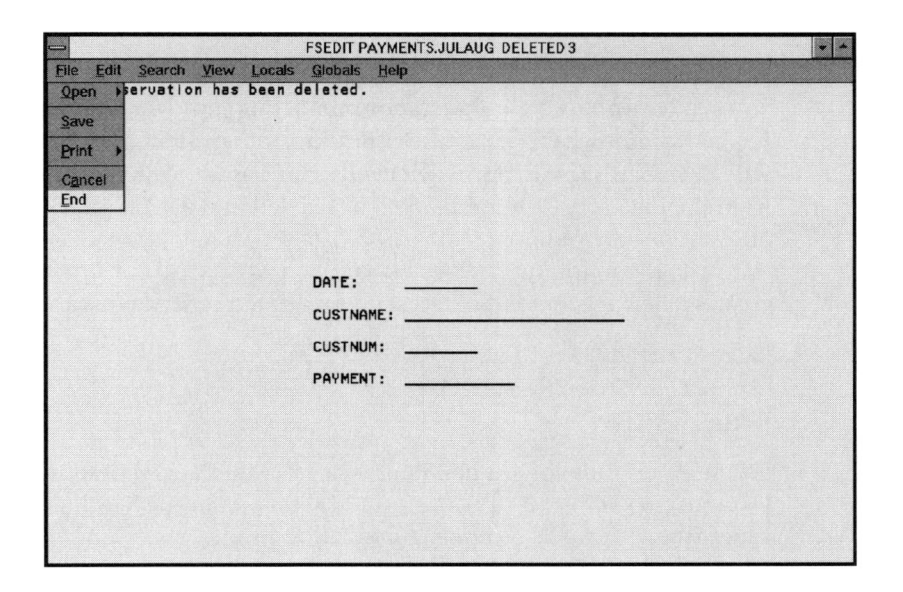

[Screenshot: FSEDIT PAYMENTS.JULAUG DELETED 3 with File menu showing Open, Save, Print, Cancel, End. Message: "servation has been deleted." with blank DATE, CUSTNAME, CUSTNUM, PAYMENT fields]

When you select **Delete record**, all values in the current observation are set to missing and the observation is marked as deleted.

You have now completed making changes to PAYMENTS.JULAUG. Select **File** from the action bar, then select **End** to save the changes and exit PROC FSEDIT.

Chapter 3 - Interactive Editing 39

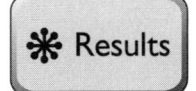

✱ Results | Updated Payments Data Set

Below is a PROC PRINT listing of the PAYMENTS.JULAUG data set after the FSEDIT session. The SUNTECH SOLAR payment (observation 5) is corrected, the payment for PAGLIA LANDSCAPE has been added (observation 12), and HANSEN DEVELOPMENT is deleted.

Updated SAS data set PAYMENTS.JULAUG

OBS	DATE	CUSTNUM	CUSTNAME	PAYMENT
1	08/12/93	83020	EXCLUSIVE PRINTS	1340.32
2	07/01/93	91109	SOUTHEAST SUPPLY	7645.01
4	08/11/93	87811	B & B DESIGN	765.00
5	08/26/93	88028	SUNTECH SOLAR	122.37
6	08/21/93	90016	HOLMES ASSOC.	5010.96
7	07/30/93	93719	CHAVEZ PUBLISHING	2654.25
8	08/31/93	89953	THE HOLWELL GROUP	10276.45
9	08/22/93	85611	EL CENTRO SERVICES	652.00
10	08/11/93	88338	BAKER CREATIVE	1246.46
11	07/06/93	90901	SHELLY & KEATS INC.	912.10
12	08/28/93	90991	PAGLIA LANDSCAPE	454.21

🔍 Closer Look

You may have noticed the **observation numbering gap** in the updated PAYMENTS.JULAUG data set. It goes from 2 to 4. Observation 3 is missing! Observation 3 was the deleted payment for HANSEN DEVELOPMENT. When you delete observations with FSEDIT, they are not immediately removed from the data set. They are marked as deleted and are unavailable for most operations. Deleted observations are removed from a SAS data set when it is re-created in a DATA or PROC step. For example, the following DATA step simply reads PAYMENTS.JULAUG with a SET statement and creates a new version of the data set. Any deleted observations are left behind in the process.

```
DATA PAYMENTS.JULAUG;
    SET PAYMENTS.JULAUG;
RUN;
```

For more information on handling data sets with deleted observations, see "Managing an Edited Data Set" in Chapter 3, "Browsing and Editing SAS Data Sets Using the FSBROWSE and FSEDIT Procedures," *SAS/FSP Software: Usage and Reference.*

Entering Dates, Finding Data Values

Entering dates. In the PAYMENTS.JULAUG data set, the variable DATE is a SAS date, that is, a numeric variable representing the number of days since Jan. 1, 1960. The DATE value for the new PAGLIA LANDSCAPE observation is entered as the characters: "8/28/93". Normally, PROC FSEDIT will not let you enter non-numeric characters like "/" in a numeric variable like DATE. It works for the PAYMENTS.JULAUG data set because the DATE variable has been assigned the MMDDYY8. informat. Once an informat is assigned, you must enter data in a form that complies with that informat. For example, you could not enter "22JUL93" or "JULY 22, 1993" in the DATE field because these do not comply with the MMDDYY8. informat. Informats also provide a measure of data validation. For example, if you try to enter "8/32/93" it will be flagged as an invalid date. If you do not assign informats to date values, you have to enter the exact number of days since Jan. 1, 1960!

There are several ways to assign an informat to a SAS variable. You can use the DATA step, PROC FSEDIT (for new data sets), and PROC DATASETS (for existing data sets). See Chapter 13, "SAS Informats" and "INFORMAT" in Chapter 9, "SAS Language Statements," *SAS Language: Reference, Version 6, First Edition;* also Chapter 17, "The DATASETS Procedure," *SAS Procedures Guide, Version 6, Third Edition;* and "Creating a New SAS Data Set" in Chapter 3, "Browsing and Editing SAS Data Sets Using the FSBROWSE and FSEDIT Procedures," *SAS/FSP Software: Usage and Reference.*

Finding data values. There are five ways to search for observations with PROC FSEDIT:

- STRING and SEARCH commands, used in this chapter, work only on character variables.

- NAME and LOCATE commands find exact matches for character and numeric variables.

- The FIND command allows you to enter logical expressions such as: "FIND PAYMENT>1000" to find all observations with PAYMENT values over $1,000.

- The WHERE command allows you to select a subset of all observations in the data set.

- You can go to a specific observation by entering the observation number on the command line.

See "Locating Observations by Content" and "Using WHERE Clauses" in Chapter 3, "Browsing and Editing SAS Data Sets Using the FSBROWSE and FSEDIT Procedures," *SAS/FSP Software: Usage and Reference,* for examples of these search methods.

Interactive editing

1. To use the FSEDIT procedure you must have SAS/FSP software installed on your system and be at a compatible workstation, PC, or terminal.

2. Make sure you know the name and have access to the SAS data set with which you want to work. This may require a LIBNAME statement. (See Chapter 2 for more information on LIBNAME.)

3. Start PROC FSEDIT with your data set named as input in the DATA= option.

4. Use the STRING and SEARCH commands to find text in character variables. FIND, NAME and LOCATE, and WHERE commands work with both character and numeric variables.

5. Create or delete observations with the Edit/Add and Edit/Delete commands.

6. The edited SAS data set is automatically saved when you exit.

FSEDIT Procedure

PROC FSEDIT general reference
   Chapter 13, "The FSEDIT Procedure," *SAS/FSP Software: Usage and Reference, Version 6, First Edition*
PROC FSEDIT overview and examples
   Chapter 3, "Browsing and Editing SAS Data Sets Using the FSBROWSE and FSEDIT Procedures," *SAS/FSP Software: Usage and Reference*

Chapter 4
Sorting and Grouping Data

Learn How to...

- Sort data

- Handle data in groups

- Print a grouped (control break) report with totals

- Select data for inclusion in a report

- Print data with dollar signs and commas

Using These SAS System Features...

- SORT procedure

- BY statement

- PRINT procedure, SUM and VAR statements

- FORMAT statement

❓ Problem Create a Sales Report Grouped by Region

You need to produce a quarterly sales report that lists sales for each marketing region and shows the regional and total sales. Individual sales are recorded in the SAS data set SALEDEPT.SALES, shown below. DATE is a SAS date variable shown in the DATE7. format. CUSTNUM is the customer number. REGION is the sales region, and AMOUNT is the amount of the sale. Only the marketing region, customer number, and sale amount should appear on the report you create.

The SALEDEPT.SALES data set

OBS	DATE	CUSTNUM	REGION	AMOUNT
1	11FEB93	991234	WEST	1733.33
2	22JAN93	967456	WEST	1304.50
3	09FEB93	058111	EAST	1086.54
4	15FEB93	991234	WEST	2245.93
5	19FEB93	958111	WEST	750.12
6	05FEB93	910023	WEST	809.66
7	05JAN93	030912	EAST	2141.89
8	25JAN93	091234	EAST	934.81
9	02MAR93	058111	EAST	1930.83
10	10MAR93	010023	EAST	2505.77
11	19JAN93	922010	WEST	1996.44
12	07FEB93	078810	EAST	769.48
13	31DEC92	991234	WEST	1580.69
14	22MAR93	978810	WEST	1517.64
15	14JAN93	958111	WEST	543.76
16	12JAN93	991234	WEST	2548.00
17	07FEB93	922010	WEST	1655.32
18	30MAR93	091234	EAST	2460.98
19	26MAR93	910023	WEST	1356.79
20	20MAR93	091234	EAST	524.77
21	26JAN93	067456	EAST	386.67
22	03JAN93	922010	WEST	915.82
23	10FEB93	991234	WEST	936.08
24	30JAN93	091234	EAST	1218.94
25	12MAR93	022010	EAST	691.98
26	26FEB93	991234	WEST	2927.13
27	27MAR93	991234	WEST	1626.77
28	15FEB93	991234	WEST	1002.52
29	19MAR93	910023	WEST	2608.15
30	25FEB93	911003	WEST	1244.20

Solution: PROC SORT, PROC PRINT, BY Statement

To generate the regional sales report you must do two things: (1) group the data by region, and (2) sum the sale amount for each region. Program 4 shows how to do this with SAS procedures SORT and PRINT. Using the VAR statement in PROC PRINT, you select the CUSTNUM and AMOUNT variables for inclusion in the report. Add a title with the TITLE statement.

Program 4

```
1   PROC SORT DATA=SALEDEPT.SALES;
2      BY REGION AMOUNT;
3   RUN;
4
5   TITLE 'Regional Sales';
6
7   PROC PRINT DATA=SALEDEPT.SALES;
8      BY REGION;
9      FORMAT AMOUNT DOLLAR11.2;
10     VAR CUSTNUM AMOUNT;
11     SUM AMOUNT;
12  RUN;
```

Program 4 Notes

Line

1 The PROC statement begins the SORT procedure. The SAS data set to be sorted is named in the DATA= option. The original version of SALEDEPT.SALES will be replaced by the sorted version. "SALEDEPT" is a SAS libref. It refers to a permanent SAS data library. It is assumed that the SALEDEPT libref has been established for you or that you have used a LIBNAME statement to create it before running the program. See Chapter 2 for an example of the LIBNAME statement, or see "LIBNAME" in Chapter 9, "SAS Language Statements," *SAS Language: Reference, Version 6, First Edition.*

2 The BY statement is required with PROC SORT. It names the sort variables. You can name one or more variables. The data set is sorted by the far-left variable first. Here the data set is sorted first by REGION, then by AMOUNT within each REGION.

3 The RUN statement ends and executes the PROC SORT step.

5 The TITLE statement will apply to the following PRINT procedure. The quoted text will appear at the top of each page of output.

7 The PROC statement begins the PRINT procedure. The DATA= option names the data set to print, in this case, the newly sorted version of SALEDEPT.SALES.

8 The BY statement tells PROC PRINT to group the data by region. The input SAS data set must be sorted by REGION for this to work.

9 The FORMAT statement tells PROC PRINT to display the values of AMOUNT with dollar signs, commas, and two decimal places. "11." means a maximum of 11 column spaces is allowed, enough for values up to $999,999.99.

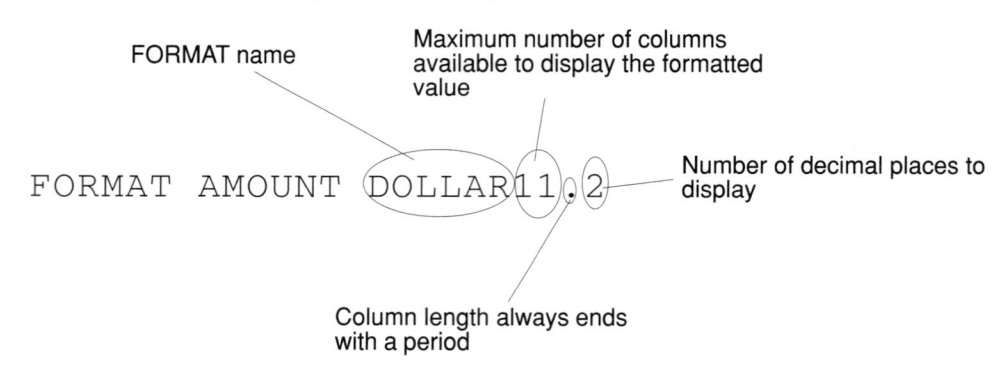

```
                                 Maximum number of columns
          FORMAT name             available to display the formatted
                                  value

                                                       Number of decimal places to
   FORMAT  AMOUNT  DOLLAR11.2                          display


                          Column length always ends
                          with a period
```

10 The VAR statement tells PROC PRINT which variables to print and the order of the columns. The CUSTNUM and AMOUNT variables appear as columns on the report. The variable DATE is in the SALEDEPT.SALES data set but is not printed because it is not named in the VAR statement. REGION appears on the report because BY variables are always printed.

11 The SUM statement tells PROC PRINT to calculate the total of AMOUNT. Because there is also a BY statement, AMOUNT is totaled for each BY group, then overall.

12 The RUN statement ends and executes the PROC PRINT step.

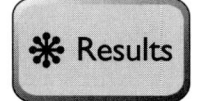

✱ Results ## Regional Sales Reports

The output on the following page shows the regional sales report. When you include a BY statement in the PROC PRINT step, it places a header in front of each BY group. In this header the name of the BY variable is followed by an equal sign and the value of the variable for the following data group. As a result of the SUM statement (line 11), AMOUNT is totaled for each BY group and overall. Note that because the data set was sorted by AMOUNT within REGION, sale amounts are listed in ascending order in each group.

The FORMAT statement (line 9) causes the AMOUNT variable to be displayed with dollar signs and commas. See End Notes for an example of an alternative PROC PRINT output format.

🔍 Closer Look Several **options are available with PROC PRINT**. These options control the listing of observation numbers (NOOBS), printing the total number of observations (N), column headings (LABEL, SPLIT=), page layout (DOUBLE, UNIFORM), and numeric precision (ROUND). There is also a PAGEBY statement that allows you to place grouped data on separate pages. See Chapter 27, "The PRINT Procedure," *SAS Procedures Guide, Version 6, Third Edition* for more information.

The regional sales report, using the BY statement

```
                            Regional Sales

    --------------------------- REGION=EAST ------------------------------

                  OBS       CUSTNUM           AMOUNT

                   1        067456           $386.67
                   2        091234           $524.77
                   3        022010           $691.98
                   4        078810           $769.48
                   5        091234           $934.81
                   6        058111         $1,086.54
                   7        091234         $1,218.94
                   8        058111         $1,930.83
                   9        030912         $2,141.89
                  10        091234         $2,460.98
                  11        010023         $2,505.77
                                            ----------
               REGION                       $14,652.66


    --------------------------- REGION=WEST ------------------------------

                  OBS       CUSTNUM           AMOUNT

                  12        958111           $543.76
                  13        958111           $750.12
                  14        910023           $809.66
                  15        922010           $915.82
                  16        991234           $936.08
                  17        991234         $1,002.52
                  18        911003         $1,244.20
                  19        967456         $1,304.50
                  20        910023         $1,356.79
                  21        978810         $1,517.64
                  22        991234         $1,580.69
                  23        991234         $1,626.77
                  24        922010         $1,655.32
                  25        991234         $1,733.33
                  26        922010         $1,996.44
                  27        991234         $2,245.93
                  28        991234         $2,548.00
                  29        910023         $2,608.15
                  30        991234         $2,927.13
                                            ----------
               REGION                       $29,302.85
                                            ==========
                                            $43,955.51
```

More on Sorting, PROC PRINT, and Formats

Sorting. Sorting is probably the most common data processing task. You may want to sort data so they can be displayed in a certain sequence, for example a report of all sales in the past year listed from the smallest to the largest. In the SAS System, sorting serves another important function: grouping data.

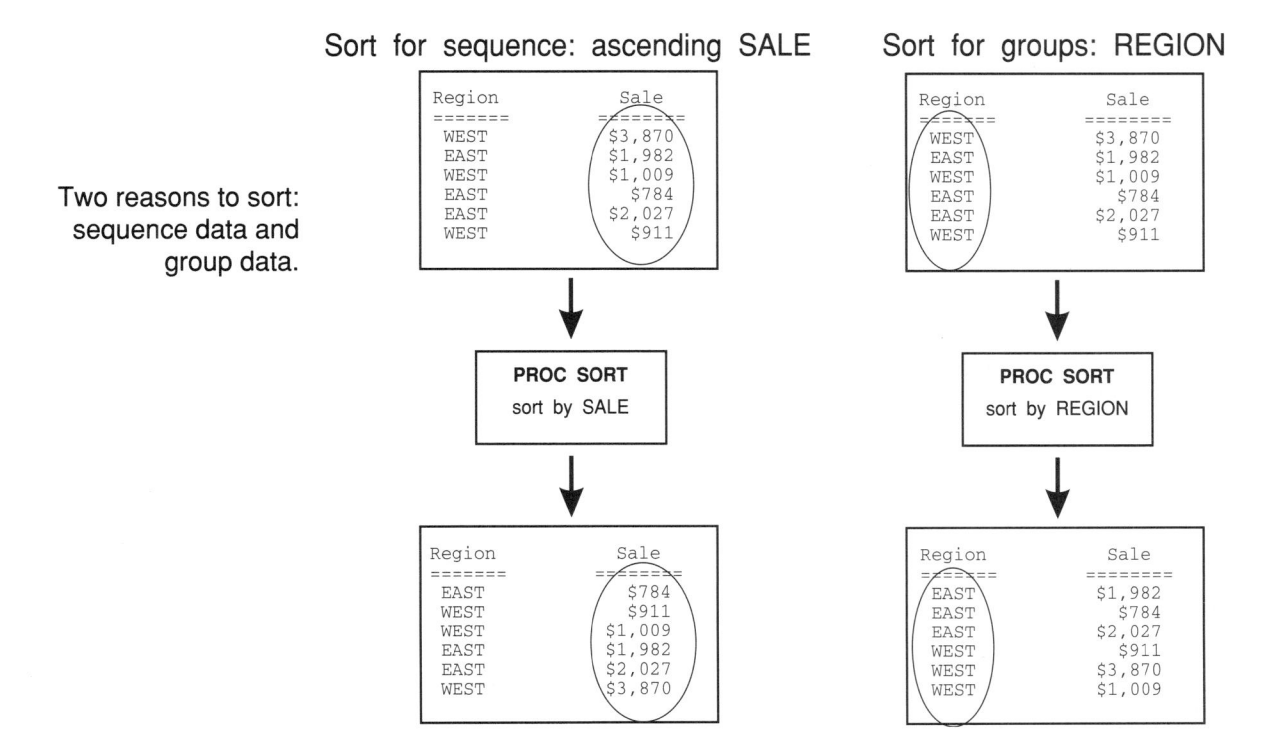

Two reasons to sort: sequence data and group data.

Sort for sequence: ascending SALE

Region	Sale
WEST	$3,870
EAST	$1,982
WEST	$1,009
EAST	$784
EAST	$2,027
WEST	$911

↓ PROC SORT sort by SALE ↓

Region	Sale
EAST	$784
WEST	$911
WEST	$1,009
EAST	$1,982
EAST	$2,027
WEST	$3,870

Sort for groups: REGION

Region	Sale
WEST	$3,870
EAST	$1,982
WEST	$1,009
EAST	$784
EAST	$2,027
WEST	$911

↓ PROC SORT sort by REGION ↓

Region	Sale
EAST	$1,982
EAST	$784
EAST	$2,027
WEST	$911
WEST	$3,870
WEST	$1,009

Like PROC PRINT in Program 4, many SAS procedures process data in groups when you include a BY statement. When you use a BY statement with a SAS procedure, you must ensure that the input data set is in BY variable order. If the data set is not already in proper order, use PROC SORT to sort it.

Preserving the original data set. When you use PROC SORT you can replace the original data set with the sorted version or create a new, sorted data set. To preserve the original SALEDEPT.SALES data set used in Program 4 and create a new sorted data set called SALEDEPT.BYREGION, use the OUT= option as follows:

```
PROC SORT DATA=SALEDEPT.SALES OUT=SALEDEPT.BYREGION;
    BY REGION AMOUNT;
RUN;
```

Other PROC SORT options allow you to sort in descending order (the default is ascending order) and control duplicates. See Chapter 31, "The SORT Procedure," *SAS Procedures Guide*.

PROC PRINT alternative output format. When you name a variable in a BY statement and an ID statement, PROC PRINT uses an alternative output format. As an example, a modified version of Program 4 and the resulting output is shown below. Instead of a header, each BY variable value is printed in the far-left column. The BY variable value is printed only when it changes. Many people prefer this format. The important thing is to know that you have a choice.

```
PROC SORT DATA=SALEDEPT.SALES;
   BY REGION AMOUNT;
RUN;

TITLE 'Regional Sales';

PROC PRINT DATA=SALEDEPT.SALES;
   FORMAT AMOUNT DOLLAR11.2;
   BY REGION;
   ID REGION;
   VAR CUSTNUM AMOUNT;
   SUM AMOUNT;
RUN;
```

PROC PRINT output with the ID and BY statements

```
                     Regional Sales

         REGION     CUSTNUM          AMOUNT

         EAST       067456          $386.67
                    091234          $524.77
                    022010          $691.98
                    078810          $769.48
                    091234          $934.81
                    058111        $1,086.54
                    091234        $1,218.94
                    058111        $1,930.83
                    030912        $2,141.89
                    091234        $2,460.98
                    010023        $2,505.77
         ------                 ----------
         EAST                     $14,652.66

         WEST       958111          $543.76
                    958111          $750.12
                    910023          $809.66
                    922010          $915.82
                    991234          $936.08
                    991234        $1,002.52
                    911003        $1,244.20
                    967456        $1,304.50
                    910023        $1,356.79
                    978810        $1,517.64
                    991234        $1,580.69
                    991234        $1,626.77
                    922010        $1,655.32
                    991234        $1,733.33
                    922010        $1,996.44
                    991234        $2,245.93
                    991234        $2,548.00
                    910023        $2,608.15
                    991234        $2,927.13
         ------                 ----------
         WEST                     $29,302.85
                                ===========
                                  $43,955.51
```

SAS formats. In Program 4, the FORMAT statement applies the DOLLAR7. format to the AMOUNT variable. DOLLAR7. is an example of a numeric variable format. Other numeric formats include:

PERCENT*w.d*	- displays numbers as percentages.
WORDS*w.*	- displays numbers as words.
Z*w.d*	- displays leading zeroes.

The *w.d* stands for width and decimal places. *w.* is the total number of spaces reserved to display the formatted value, including all decimal places, dollar signs, commas, etc. *d* is the number of decimal places to display. If you do not specify *d*, no decimal places are displayed and values are rounded to whole numbers.

SAS dates are stored in numeric variables. Date formats are a special class of numeric formats. The value of a date variable represents the number of days between January 1, 1960 and the given date value. Date formats translate this number into familiar date layouts. The values of the variable DATE shown in the Problem section of this chapter are displayed in the DATE7. format. Some other date formats include:

MMDDYY*w.*	- month, day, year ("07/04/93")
YYMMDD*w.*	- year, month, day ("93/07/04")
MONYY*w.*	- month, year ("JUL93")

In addition to numeric formats, there are several character formats, for example:

$HEX*w.*	- displays values as hexadecimal strings.
$VARYING*w.*	- displays values in varying display widths.
$*w.*	- displays the default character format.

Note that character format names all begin with "$". As with numeric formats, *w.* determines the display width, or, in the case of the $VARYING format, the maximum possible width.

It is usually not necessary to specify a format for a character variable. The default character display format ($*w.*) is appropriate in most cases.

Sorting and Grouping Data

1. Make sure you have access to the SAS data set you want to sort. Use a LIBNAME statement if using a permanent SAS data set. (See Chapter 2 in this book for more information on permanent SAS data sets.)

2. Use a PROC SORT step that names the data set to be sorted in the DATA= option. Use the OUT= option if you do not want to replace the original data set with the sorted version (see End Notes).

3. Include a BY statement that names the sort variables.

4. Use a BY statement in subsequent SAS procedures such as PROC PRINT to process data in groups.

Sorting, Printing, and Formats

FORMAT statement
  "FORMAT" in Chapter 9, "SAS Language Statements," *SAS Language: Reference, Version 6, First Edition*
PRINT procedure, BY statement, ID statement, VAR statement
  Chapter 27, "The PRINT Procedure," *SAS Procedures Guide, Version 6, Third Edition*
SAS data set indexes
  "Release 6.06: SAS Indexes" in Chapter 6, "SAS Files," *SAS Language: Reference*
SAS formats
  "SAS Formats" in Chapter 3, "Components of the SAS Language," *SAS Language: Reference*
SORT procedure, BY statement, OUT= option
  Chapter 31, "The SORT Procedure," *SAS Procedures Guide*

Part 2 - Working with Your Data

Chapter 5 -   Counting Occurrences

Chapter 6 -   Finding Sums

Chapter 7 -   Finding Averages

Chapter 8 -   Grouping Data: Days into Months

Chapter 9 -   Grouping Data: Create Your Own Data Groups

Chapter 10 - Combining SAS Data Sets

Chapter 11 - Selecting and Changing Data

Chapter 12 - Finding Unique Values

Chapter 13 - Finding Percentages

Chapter 5
Counting Occurrences

Learn How to...

- Count occurrences of specific values
- Produce an occurrence or frequency report
- Create a data set of frequency information

Using These SAS System Features...

- FREQ procedure
- TABLES statement
- OUT= option

? Problem — Create a Quality Control Report

A precision molding company has three production lines that run around the clock. Each line has a quality check station that measures parts to ensure that critical dimensions are within design tolerances. When a part does not pass the quality check, it is rejected to a recycling bin. For each reject, the quality check station automatically records the hour of the day (0 to 23) and the production line number on a diskette in the monitoring station computer. Each entry in this file represents one rejected part. You can use a SAS DATA step program to read the rejects diskette from each production line and create the SAS data set QCDATA.REJECTS, shown below. HOUR is the time of day and LINE is the production line number.

You need to generate a report from the rejects file that shows the total number of rejects for each line during the 24-hour period.

SAS data set QCDATA.REJECTS

OBS	HOUR	LINE
1	0	1
2	0	2
3	1	3
4	2	3
5	3	1
6	3	2
7	3	3
8	4	3
9	5	1
10	8	1
11	9	3
12	10	1
13	10	2
14	10	3
15	11	2
16	11	3
17	12	3
18	13	1
19	13	3
20	14	1
21	15	2
22	15	3
23	16	2
24	17	1
25	18	1
26	18	3
27	19	2
28	19	3
29	20	1
30	20	2
31	22	1
32	22	2
33	23	3

Solution: FREQ Procedure

Because each observation in the QCDATA.REJECTS data set represents a single rejected part for a molding line, you can solve this problem by counting the number of observations for each value of the variable LINE. Counting occurrences of unique values of a variable is called frequency analysis—you are determining the frequency of each unique value. The easiest way to do this in the SAS System is with PROC FREQ. To use PROC FREQ you need two pieces of information: the name of the SAS data set holding the information, and the name of the variable you want to examine for unique values.

Name the input data set in the DATA= option of the PROC FREQ statement. You name the variable you want to look at in a TABLE statement. You can generate a frequency report directly with PROC FREQ, as shown in Program 5 below, or you can create a SAS data set that contains the same information, as shown in End Notes. Either way, PROC FREQ is an easy-to-use yet powerful data analysis tool.

Program 5

```
1  PROC FREQ DATA=QCDATA.REJECTS;
2     TITLE 'Mold Reject Count';
3     TABLE LINE;
4  RUN;
```

Program 5 Notes

Line 1 The PROC statement starts PROC FREQ and names the SAS data set to analyze: QCDATA.REJECTS.

2 The TITLE statement applies to the PROC FREQ output. The quoted text will appear at the top of each page. See Closer Look below for a discussion of TITLE statement placement.

3 The TABLE statement names the variable you want to analyze: LINE.

4 The RUN statement ends and executes the PROC FREQ step.

Closer Look

The **TITLE statement** belongs to a family of SAS statements called global statements. Such statements are independent of any one PROC or DATA step. Once title text is assigned with a TITLE statement, it is in effect for all printed or displayed output that follows, until you execute a new TITLE statement to replace it.

Another feature of global statements is that they may appear anywhere. The TITLE statement in Program 5 could have been placed before the PROC FREQ step:

```
TITLE 'Mold Reject Count';
PROC FREQ DATA=QCDATA.REJECTS;
    TABLE LINE;
RUN;
```

When you place a global statement inside a step, it applies to that step and all subsequent steps. When you place a global statement outside a step it applies to the next step and all subsequent steps. The TITLE statement, like all global statements, does not work backwards; that is, the statement does not apply to output from steps that appear before it in your program.

❋ Results

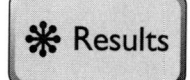

Rejected Parts Report

Program 5 produces the report shown below. Each of the three production lines is represented in the `LINE` column. The `Frequency` column contains the reject counts for each line. The numbers in this column represent the number of observations in the QCDATA.REJECTS data set for each value of `LINE`. This is the total reject count for each line.

Other columns are part of the default output from PROC FREQ. `Percent` shows the percent of all rejects accounted for by each line. `Cumulative Frequency` shows the number of rejects for the current line plus all previous lines. `Cumulative Percent` is the percent accounted for by the current plus all previous lines.

Rejected parts report

```
                         Mold Reject Count

                                         Cumulative   Cumulative
        LINE    Frequency    Percent     Frequency     Percent
        ---------------------------------------------------------
          1          11        33.3          11          33.3
          2           9        27.3          20          60.6
          3          13        39.4          33         100.0
```

🔍 Closer Look

There are several options you can use with **PROC FREQ to control output**. In the PROC statement you can specify the reporting order with the ORDER= option. By default, data appear in "internal" order, which means by ascending value of the TABLE statement variable. You can also specify that data be listed in descending frequency order, the order in which data are encountered in the input data set, or the order of the formatted value of the TABLE variable.

TABLE statement options allow you to suppress the cumulative and percentage data. PROC FREQ also has the capability to generate crosstabulations that show counts for each combination of two or more variables. See Chapter 20, "The FREQ Procedure," in *SAS Procedures Guide, Version 6, Third Edition*.

Counting, PROC FREQ Output Data Sets

Other ways to count. There are a number of SAS procedures that can do frequency analysis, that is, count unique values of variables. In base SAS software PROC MEANS, PROC SUMMARY, and PROC TABULATE generate frequency counts. You can use one of these instead of PROC FREQ if you need to generate additional statistics such as the mean and standard deviation. Also note that you can use PROC PRINT to generate frequency counts. In the SAS System, frequency is referred to as the N statistic. Use the N option with PROC PRINT and a BY statement that names the variable you want to examine as follows:

```
PROC PRINT DATA=SOMELIB.SOMEDATA N;
    BY SOMEVAR;
RUN;
```

PROC PRINT prints all the observations for each unique value of SOMEVAR then reports the count of those observations, that is, the frequency of each unique value of SOMEVAR. See Chapter 4 in this book for an example of the BY statement with PROC PRINT. Also see Chapter 21, "The MEANS Procedure," Chapter 27, "The PRINT Procedure," Chapter 36, "The SUMMARY Procedure," and Chapter 37, "The TABULATE Procedure," in the *SAS Procedures Guide*.

Counting discrete and continuous variables. Frequency counts are normally done on discrete variables. Discrete variables have a limited set of values. For example, a variable that has a set of values that consists of a single letter of the alphabet could only have one of 26 possible values. By contrast, a continuous variable can hold any one of a large set of values. Consider a variable that holds the elapsed time for the mile run at a track meet. The number of possible values is limited only by the precision of the timing clock. Between finish times of 4 minutes and 4 minutes 10 seconds, there are 100 possible values if timed to the tenth of a second, but 1000 possible values if timed to the hundredth of a second.

Direct frequency analysis of continuous variables is usually not very meaningful. In most cases it is more useful to establish ranges of values, then count occurences that fall within each range. For example, mile run times could be grouped as all times 4 minutes or above, and all times below 4 minutes. The set of possible values is reduced from thousands to just two: below and above 4 minutes. There are various ways to group continuous variables into discrete categories. See Chapter 11 in this book for an example.

PROC FREQ output data set. In Program 5, PROC FREQ creates printed output. By adding the OUT= option to the TABLE statement, you can use PROC FREQ to generate a SAS data set containing the frequency information. In the program below, PROC FREQ creates a reject count data set called QCDATA.LINECNT. The data set is printed by a subsequent PROC PRINT step.

```
TITLE 'Mold Reject Count';
PROC FREQ DATA=QCDATA.REJECTS;
   TABLE LINE / OUT=QCDATA.LINECNT NOPRINT;
RUN;
PROC PRINT DATA=QCDATA.LINECNT;
RUN;
```

Output from the PROC PRINT step

```
               Mold Reject Count

     OBS     LINE     COUNT     PERCENT

      1        1        11      33.3333
      2        2         9      27.2727
      3        3        13      39.3939
```

Note that TABLE statement options must be preceded by a "/". Because of the NOPRINT option in the TABLE statement, the FREQ procedure above does not generate a printout. The output is the result of the PRINT procedure, not the PROC FREQ step.

The program above demonstrates an alternative way to use the FREQ procedure: generate a SAS data set holding the frequency information, then process that data set with another SAS procedure. The SAS data set QCDATA.LINECNT, generated by the PROC FREQ step, has 3 variables:

LINE -   is the TABLE statement variable.

COUNT - contains the same information as the Frequency column in the standard PROC FREQ output.

PERCENT - contains the same information as the Percent column in the standard PROC FREQ output.

Cumulative data are not included in the PROC FREQ output data set.

When you create an output dataset with the FREQ procedure, you have more flexibility with handling and presenting the data. On the other hand, if all you need is a simple printed report of frequency counts, it couldn't be much easier than using PROC FREQ as shown in Program 5.

Counting Occurrences

1. Make sure you know the name and have access to the SAS data set holding the data you want to examine. This may require a LIBNAME statement (See Chapter 2 in this book for a LIBNAME example.)

2. Make sure you know the name of the SAS variable you want to examine.

3. Write a PROC FREQ step that names your input SAS data set in the DATA= option.

4. Name the variable you want to examine in a TABLE statement in the PROC FREQ step.

5. To create an output SAS data set and suppress printed output, add TABLE statement options OUT= and NOPRINT. TABLE options must be preceded by a "/".

FREQ Procedure, Global Statements

global SAS statements
  "Global Statements" in Chapter 3, "Components of the SAS Language," *SAS Language: Reference, Version 6, First Edition*
PROC FREQ
  Chapter 20, "The FREQ Procedure," *SAS Procedures Guide, Version 6, Third Edition*
PROC FREQ output SAS data set
  "DETAILS, Output Data Set" in Chapter 20, "The FREQ Procedure," *SAS Procedures Guide*
TABLES statement and options
  "TABLES Statement" in Chapter 20, "The FREQ Procedure," *SAS Procedures Guide*

Chapter 6
Finding Sums

Learn How to...

- Create data groups without sorting
- Generate a report with group totals
- Create a data set of summary information

Using These SAS System Features...

- MEANS procedure
- CLASS statement
- OUTPUT statement

Problem: Generate an Equipment Purchases Report

The accounting department tracks computer equipment purchases in a SAS data set named ACCOUNT.EQUIPMNT shown below. Every time a piece of equipment is purchased the name of the department that made the purchase is recorded in the variable DEPT, the type of equipment in variable EQUIP, and the cost in AMOUNT. You need to create a report showing the total spent by each department for each type of equipment.

SAS data set ACCOUNT.EQUIPMNT

OBS	DEPT	EQUIP	AMOUNT
1	MARKET	PRINTER	1933.55
2	MARKET	PC	2485.17
3	MARKET	DISK	1431.80
4	MARKET	PRINTER	2372.60
5	MARKET	PC	3960.88
6	MARKET	DISK	1165.10
7	MARKET	MEMORY	357.00
8	MANUFAC	PC	2056.59
9	MANUFAC	DISK	504.94
10	ADMIN	PC	2133.54
11	ADMIN	DISK	693.33
12	ADMIN	MEMORY	435.00
13	MARKET	PRINTER	1295.51
14	MARKET	PC	2264.53
15	ADMIN	PRINTER	1462.26
16	ADMIN	DISK	927.08
17	ADMIN	MEMORY	353.00
18	MANUFAC	PRINTER	2331.88
19	MANUFAC	PC	3048.02
20	MANUFAC	MEMORY	544.00
21	MANUFAC	PRINTER	1609.84
22	MANUFAC	PC	5849.14
23	MANUFAC	DISK	541.68
24	MARKET	PRINTER	823.54
25	MARKET	PC	4365.36
26	MARKET	DISK	677.05
27	MARKET	MEMORY	340.00
28	MARKET	PRINTER	2451.09
29	MARKET	DISK	1442.65
30	MARKET	MEMORY	394.00
31	ADMIN	PRINTER	1593.22
32	ADMIN	PC	4697.41
33	ADMIN	DISK	944.09
34	ADMIN	MEMORY	397.00
35	MANUFAC	PC	5944.24
36	MANUFAC	DISK	1281.90
37	MANUFAC	MEMORY	488.00

MEANS Procedure

To generate the equipment purchase report, you need to group by department and equipment type, then sum the purchases within each of these categories. Any time you have a problem that calls for analyzing data groups, you should consider PROC MEANS. The CLASS statement in PROC MEANS allows you to create data categories or groups without sorting. In this case you need to group data according to the values of department and the type of equipment purchased. Program 6 shows how to create the report with just five SAS statements.

This example calls on PROC MEANS to produce the sum statistic, that is, add things up. Many other statistics are available including mean (average), minimum, maximum, range, and N (the count of nonmissing values). Like PROC FREQ in Chapter 5, PROC MEANS can create an output data set in addition to, or instead of, a printed report. See End Notes for an alternative solution to the equipment report problem using the output data set option.

Program 6

```
1   PROC MEANS DATA=ACCOUNT.EQUIPMNT SUM;
2      TITLE 'Departmental Equipment Purchases';
3      CLASS DEPT EQUIP;
4      VAR AMOUNT;
5   RUN;
```

Program 6 Notes

Line 1 The PROC statement starts the MEANS procedure. ACCOUNT.EQUIPMNT is named as the input SAS data set, and the SUM option tells PROC MEANS to generate totals.

2 The TITLE statement applies to the PROC MEANS output. The quoted text will be printed at the top of each page.

3 The CLASS statement tells PROC MEANS how to group the data. Here it groups each equipment type within each department, that is, each value of EQUIP within each value of DEPT. The group hierarchy is determined by the left-to-right order of the variables on the CLASS statement. The data set does not have to be sorted.

4 The VAR statement names the variable you want to analyze, in this case AMOUNT. You can name one or more numeric variables in a VAR statement.

5 The RUN statement ends and executes the PROC MEANS step.

Results

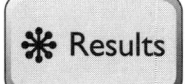

Equipment Purchases Report

The output below shows the equipment purchases report generated by PROC MEANS. The layout corresponds to the hierarchy of variables listed in the CLASS statement (line 3): EQUIP grouped within DEPT. The N Obs column shows the count of observations for each DEPT/EQUIP combination. For example, there were three disk purchases in the administration department. The final column, Sum, is the statistic requested in the PROC statement (line 1). You can see that the three administration department disk purchases added up to $2,564.50. The TITLE statement text appears at the top of each page, and PROC MEANS adds the Analysis Variable : AMOUNT subtitle.

Equipment Purchases Report from PROC MEANS

```
                 Departmental Equipment Purchases

              Analysis Variable : AMOUNT


           DEPT       EQUIP     N Obs          Sum
           -------------------------------------------
           ADMIN      DISK          3       2564.50

                      MEMORY        3       1185.00

                      PC            2       6830.95

                      PRINTER       2       3055.48

           MANUFAC    DISK          3       2328.52

                      MEMORY        2       1032.00

                      PC            4      16897.99

                      PRINTER       2       3941.72

           MARKET     DISK          4       4716.60

                      MEMORY        3       1091.00

                      PC            4      13075.94

                      PRINTER       5       8876.29
           -------------------------------------------
```

Closer Look

You may want to **present the AMOUNT totals formatted with dollar signs and commas**. You cannot use the FORMAT statement to control the PROC MEANS output statistics. Adding the statement **FORMAT AMOUNT DOLLAR11.2;** to the PROC MEANS step has no effect. This is one reason you may want to create an output data set instead of letting PROC MEANS generate your report directly. You can get dollar signs and commas by adding a FORMAT statement to a PRINT procedure that prints the PROC MEANS output data set. See End Notes for an example.

Output Data Set, PROC SUMMARY, Counting

PROC MEANS output data set. PROC MEANS can create a SAS data set instead of a printed report. This is particularly useful when you want to use the summarized data as input for some other SAS step. The program below creates the equipment purchases report in two steps. First, a SAS data set called ACCOUNT.EQUIPSUM is created with PROC MEANS, then ACCOUNT.EQUIPSUM is printed using PROC PRINT with a FORMAT statement for dollar signs:

```
PROC MEANS DATA=ACCOUNT.EQUIPMNT NOPRINT;
   CLASS DEPT EQUIP;
   VAR AMOUNT;
   OUTPUT OUT=ACCOUNT.EQUIPSUM SUM=;
RUN;

TITLE 'Departmental Equipment Purchases';
PROC PRINT DATA=ACCOUNT.EQUIPSUM;
   FORMAT AMOUNT DOLLAR11.2;
RUN;
```

The following output is the result of the PROC PRINT step above:

```
                   Departmental Equipment Purchases

   OBS      DEPT       EQUIP      _TYPE_     _FREQ_         AMOUNT

    1                                0         37        $65,595.99
    2                 DISK           1         10         $9,609.62
    3                 MEMORY         1          8         $3,308.00
    4                 PC             1         10        $36,804.88
    5                 PRINTER        1          9        $15,873.49
    6      ADMIN                     2         10        $13,635.93
    7      MANUFAC                   2         11        $24,200.23
    8      MARKET                    2         16        $27,759.83
    9      ADMIN      DISK           3          3         $2,564.50
   10      ADMIN      MEMORY         3          3         $1,185.00
   11      ADMIN      PC             3          2         $6,830.95
   12      ADMIN      PRINTER        3          2         $3,055.48
   13      MANUFAC    DISK           3          3         $2,328.52
   14      MANUFAC    MEMORY         3          2         $1,032.00
   15      MANUFAC    PC             3          4        $16,897.99
   16      MANUFAC    PRINTER        3          2         $3,941.72
   17      MARKET     DISK           3          4         $4,716.60
   18      MARKET     MEMORY         3          3         $1,091.00
   19      MARKET     PC             3          4        $13,075.94
   20      MARKET     PRINTER        3          5         $8,876.29
```

When an OUTPUT statement is present in the PROC MEANS step, a SAS data set is created. This data set is named in the OUT= option. The SUM= option tells PROC MEANS to sum up values of the variable(s) named in the VAR statement. In this case, AMOUNT is summed and the result placed in a new variable, also named AMOUNT.

Notice that when you create an output data set, the statistic option (SUM=, MEAN=, MIN=, etc.) is specified in the OUTPUT statement, not in the PROC statement as in Program 6. The NOPRINT option in the PROC statement suppresses the standard PROC MEANS printed output. Without this option the PROC MEANS step would generate both a data set and a printed report.

The OUTPUT statement format demonstrates one of the alternatives for specifying statistics. In this case, the form *statistic-keyword=* is used. The keyword SUM is followed by an equal sign. When you use this format, the output data set variable inherits the name of the VAR statement variable, AMOUNT. You can explicitly name output data set variables by using the *statistic-keyword(variable-list)=name-list* format, for example: SUM(AMOUNT)=TOTCOST. This tells the system to sum up the AMOUNT variable and put the result in a new variable named TOTCOST. See "OUTPUT Statement" in Chapter 21, "The MEANS Procedure," *SAS Procedures Guide, Version 6, Third Edition*, for details on OUTPUT statement format options.

The variables DEPT, EQUIP, and AMOUNT in ACCOUNT.EQUIPSUM correspond to variables in the input data set, ACCOUNT.EQUIPMNT. In addition, two new variables are created: _TYPE_ and _FREQ_. _FREQ_ is like the N Obs column in the Program 6 output. It represents the number of observations in each DEPT/EQUIP category. The _TYPE_ variable is assigned a number, here 0 to 3, that corresponds to the source of the information in each observation. The _TYPE_=0 observation holds information from all observations in the data set. _TYPE_=1 observations hold information for each value of EQUIP. _TYPE_=2 observations hold information for each value of DEPT. The _TYPE_=3 observations are from individual DEPT/EQUIP combinations. The _TYPE_ variable allows you to identify specific levels of data summarization. Information relating to the data set overall is always assigned _TYPE_=0. For a description of how other _TYPE_ values are assigned, see "DETAILS, Observations" in Chapter 21, "The MEANS Procedure," *SAS Procedures Guide*.

Counting with PROC MEANS. The _FREQ_ column in the output on the previous page and the N Obs column in the Program 6 output represent frequencies or counts. PROC MEANS can be used in place of PROC FREQ (see Chapter 5) for many counting problems. The MEANS procedure returns a count of non-missing values for each unique combination of the CLASS statement variables. PROC FREQ does generate percentages and cumulative statistics that PROC MEANS does not. If the only statistic you need from PROC MEANS is frequency, then specify "N" in the PROC statement or the OUTPUT statement, just as you specified "SUM" in Program 6.

Summing variables within an observation. Consider the following SAS data set, called PROFITS:

```
YEAR    QTR1    QTR2    QTR3    QTR4
1991    8240    10711   11889   9902
1992    8802    9871    9940    8621
1993    10822   11744   12009   12739
```

It has three observations. In each observation the variable YEAR indicates the fiscal year for profit reporting, QTR1 is the first quarter profits, QTR2 the second quarter profits, etc.

You need to find the total profit for each fiscal year. Can you use PROC MEANS? PROC MEANS works with the values of individual variables. You could use PROC MEANS to add up all the QTR1 values for the three years, but you cannot use PROC MEANS to find the yearly totals summed from the values of different variables in a single observation. To sum across variables for each observation, you can use a DATA step with the SUM function:

```
DATA YEARTOT;
   SET PROFITS;
   YTOTAL=SUM(QTR1,QTR2,QTR3,QTR4);
RUN;
```

The output data set YEARTOT contains a new variable called YTOTAL that holds the total profits for the year. The SUM function adds up the values in the argument list and returns the total.

SUM is just one example of the many DATA step functions available. There are functions that manipulate character strings, generate random numbers, convert ZIP codes to state names, plus many others. All SAS functions work like SUM: you name the function followed by a set of arguments enclosed in parenthesis. The function processes the arguments and returns a value. See "Using SAS Functions" in Chapter 6, "Working with Numeric Variables," *SAS Language and Procedures: Usage, Version 6, First Edition*, and "SAS Functions" in Chapter 3, "Components of the SAS Language," *SAS Language: Reference, Version 6, First Edition*.

Finding Sums

1. Make sure you know the name and have access to the SAS data set holding the data you want to examine. This may require a LIBNAME statement (See Chapter 2 in this book for a LIBNAME example.)

2. Make sure you know the names of the SAS variables with which you need to work.

3. Use PROC MEANS. Decide if you can use the printed output generated directly by PROC MEANS or if you need to produce a SAS data set for further processing.

4. Name grouping variables in the CLASS statement. Use the VAR statement to name the variables you want to sum.

5. Use the OUTPUT statement if you need an output data set. In the OUTPUT statement, name the output data set with the OUT=*datasetname* option. Name the statistics you want with *statistic-keyword=* options such as SUM=.

   The PROC statement option NOPRINT suppresses PROC MEANS printed output.

PROC MEANS, SAS Functions

CLASS statement
   "CLASS Statement" in Chapter 21, "The MEANS Procedure," *SAS Procedures Guide, Version 6, Third Edition*
PROC MEANS, general
   "Computing Basic Descriptive Statistics for Variables" in Chapter 17, "Producing Summary Reports," *SAS Language and Procedures: Usage 2, Version 6, First Edition*; Chapter 21, "The MEANS Procedure," *SAS Procedures Guide*
PROC MEANS output SAS data set
   "OUTPUT Statement" and "DETAILS, Output Data Set" in Chapter 21, "The MEANS Procedure," *SAS Procedures Guide*
PROC MEANS statement options
   "PROC MEANS Statement" in Chapter 21, "The MEANS Procedure," *SAS Procedures Guide*

Chapter 7
Finding Averages

Learn How to...

- Find averages (means)

- Create an averages report with data groups

- Select data you need from the input data set

Using These SAS System Features...

- MEANS procedure

- CLASS statement

- WHERE statement

❓ Problem: Find the Average Crop Yield in Three Counties

A state agriculture experiment station is testing a new hybrid seed corn in five counties. In each county, seeds are planted in ten plots to test yields under different soil conditions. At the end of the growing season, each plot is harvested and the yield in bushels is recorded in SAS data set AGDEPT.YIELDS, shown below. COUNTY is the county where the plot is located. PLOT is the plot ID number, and YIELD, is the plot yield in bushels. You need to find the average yields for three of the five counties: Sauk, Dane, and Columbia.

SAS data set AGDEPT.YIELDS

OBS	COUNTY	PLOT	YIELD
1	SAUK	101	181
2	SAUK	102	176
3	SAUK	103	85
4	SAUK	104	106
5	SAUK	105	149
6	SAUK	106	187
7	SAUK	107	186
8	SAUK	108	162
9	SAUK	109	126
10	SAUK	110	107
11	ST. CROIX	101	87
12	ST. CROIX	102	179
13	ST. CROIX	103	96
14	ST. CROIX	104	127
15	ST. CROIX	105	154
16	ST. CROIX	106	86
17	ST. CROIX	107	107
18	ST. CROIX	108	139
19	ST. CROIX	109	148
20	ST. CROIX	110	177
21	DANE	101	162
22	DANE	102	91
23	DANE	103	166
24	DANE	104	136
25	DANE	105	97
26	DANE	106	90
27	DANE	107	91
28	DANE	108	104
29	DANE	109	91
30	DANE	110	114

SAS data set AGDEPT.YIELDS (continued)

OBS	COUNTY	PLOT	YIELD
31	COLUMBIA	101	174
32	COLUMBIA	102	189
33	COLUMBIA	103	117
34	COLUMBIA	104	164
35	COLUMBIA	105	135
36	COLUMBIA	106	139
37	COLUMBIA	107	167
38	COLUMBIA	108	136
39	COLUMBIA	109	141
40	COLUMBIA	110	167
41	CLARK	101	90
42	CLARK	102	140
43	CLARK	103	165
44	CLARK	104	118
45	CLARK	105	88
46	CLARK	106	185
47	CLARK	107	144
48	CLARK	108	123
49	CLARK	109	93
50	CLARK	110	134

Solution — # PROC MEANS and the WHERE Statement

There are two parts to this problem. First, you need to select the three counties for the report. Then, you need to find the average yield in each of those counties.

PROC MEANS can easily solve this problem. Use the WHERE statement to select the three counties. To show the average yield by county, name the variable COUNTY in the PROC MEANS CLASS statement.

Program 7

```
1  PROC MEANS DATA=AGDEPT.YIELDS MEAN MAXDEC=1;
2     TITLE 'Average County Yields';
3     WHERE COUNTY IN('SAUK', 'DANE', 'COLUMBIA');
4     CLASS COUNTY;
5     VAR YIELD;
6  RUN;
```

Closer Look — **When you want an average or mean across all observations in a SAS data set, simply leave off the CLASS statement.** This tells PROC MEANS to forget categories and look at the entire data set. For example, if you want to find the average yield for all the plots in data set AGDEPT.YIELDS, use the following PROC MEANS step:

```
PROC MEANS DATA=AGDEPT.YIELDS MEAN MAXDEC=1;
   TITLE 'Average Yield All Counties';
   VAR YIELD;
RUN;
```

Program 7 Notes

Line 1 The PROC statement starts the MEANS procedure and names the SAS data set AGDEPT.YIELDS as input. The MEAN keyword says you want PROC MEANS to calculate the mean or average. The MAXDEC=1 option limits the number of decimal places displayed in the PROC MEANS output.

2 The TITLE statement applies to the PROC MEANS output. The quoted text will be displayed at the top of each page.

3 The WHERE statement limits the observations read by the procedure to those that satisfy the WHERE expression. In this case, observations are read only where the value of the variable COUNTY is in the set of three counties listed. Since COUNTY is a character variable, you must enclose values in quotes.

4 The CLASS statement tells PROC MEANS that you want the mean reported for each unique value of COUNTY.

5 The VAR statement names the variable you want to analyze. In this case, you want to find the mean of YIELD.

6 The RUN statement ends and executes the MEANS procedure.

Closer Look — **The WHERE statement can be used in most SAS procedures.** It acts as a filter on the input SAS data set. Only observations that satisfy the WHERE expression are available to the PROC. You can use all the standard logical operators in a WHERE expression: >, <, AND, OR, =, etc. In addition, there are some WHERE-statement-only operators such as BETWEEN-AND, CONTAINS, and LIKE. These are particularly useful for selecting data based on partial matches of character variables. See "WHERE" in Chapter 9, "SAS Language Statements," *SAS Language, Reference: Version 6, First Edition* for details.

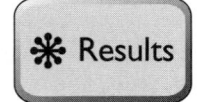

❋ Results Average Yield Report

The results of Program 7 appear below. Each unique value of the CLASS variable COUNTY is listed followed by the number of observations found for each county and the mean yield. PROC MEANS adds the subtitle "Analysis Variable : YIELD". The yield means contain one decimal place as specified by the MAXDEC=1 option (line 1).

PROC MEANS makes it very easy to find averages. Just remember that you name the value you want to average in the VAR statement and the data groups you need in the CLASS statement. PROC MEANS can generate many other statistics besides the mean and can create a SAS data set as well as the printed output shown here. See Chapters 6 and 8 in this book for more PROC MEANS examples.

Average yield report

```
                       Average County Yields

                    Analysis Variable : YIELD


                 COUNTY        N Obs            Mean
                 -----------------------------------
                 COLUMBIA         10           152.9

                 DANE             10           114.2

                 SAUK             10           146.5
                 -----------------------------------
```

MEANS Procedure Default Output

PROC MEANS default output. In this chapter and in Chapter 6, you specify the desired statistic in the PROC MEANS statement or in an OUTPUT statement. What happens if you don't specify any statistics? If you omit "MEAN" from the Program 7 PROC statement, PROC MEANS generates the following output:

PROC MEANS default output

```
                        Average County Yields

   Analysis Variable : YIELD


   COUNTY      N Obs    N          Mean        Std Dev       Minimum
   ---------------------------------------------------------------------
   COLUMBIA     10     10          152.9          22.4         117.0

   DANE         10     10          114.2          29.9          90.0

   SAUK         10     10          146.5          37.9          85.0
   ---------------------------------------------------------------------
                    COUNTY      N Obs     Maximum
                    --------------------------------
                    COLUMBIA     10        189.0

                    DANE         10        166.0

                    SAUK         10        187.0
                    --------------------------------
```

The default PROC MEANS statistics are the number of observations read (N Obs), the number of observations with nonmissing values (N), the mean, standard deviation, minimum, and maximum.

If you create a SAS data set with PROC MEANS but omit any statistics keywords from the OUTPUT statement, the resulting data set will contain the N, minimum, maximum, mean, and standard deviation. Each statistic will be identified in a variable called _STAT_. For more details see "Creating the Default _STAT_ Data Set" in Chapter 21, "The MEANS Procedure," *SAS Procedures Guide, Version 6, Third Edition*.

Finding Averages

1. Make sure you know the name and have access to the SAS data set holding the data you want to examine. This may require a LIBNAME statement. (See Chapter 2 in this book for a LIBNAME example.)

2. Make sure you know the names of the SAS variables with which you need to work.

3. Use PROC MEANS with the MEAN keyword specified in the PROC statement. Use the MAXDEC= option to limit the number of decimal places displayed.

4. If you need to filter data, use the WHERE statement.

5. Use the CLASS statement to group data and the VAR statement to name the variable you want averaged.

Using the MEANS Procedure

CLASS statement
  "CLASS Statement" in Chapter 21, "The MEANS Procedure," *SAS Procedures Guide, Version 6, First Edition*
PROC MEANS, general
  "Computing Basic Descriptive Statistics for Variables" in Chapter 17, "Producing Summary Reports Using Descriptive Procedures," *SAS Language and Procedures: Usage 2, Version 6, First Edition;*
  Chapter 21, "The MEANS Procedure," *SAS Procedures Guide*
PROC MEANS statement options
  "PROC MEANS Statement" in Chapter 21, "The MEANS Procedure,"*SAS Procedures Guide*
statistics available
  "Statistic Keywords Available in the PROC MEANS Statement" in Chapter 21, "The MEANS Procedure," *SAS Procedures Guide*
WHERE statement
  "WHERE" in Chapter 9, "SAS Language Statements," *SAS Language: Reference, Version 6, First Edition*

Chapter 8
Grouping Data: Days into Months

Learn How to...

- Create data groups

- Group daily data by month

- Create a grouped summary report

Using These SAS System Features...

- FORMAT statement

- SAS date formats

- MEANS Procedure

? Problem — Find the Monthly Total Rainfall

The local weather station records daily precipitation in the SAS data set WEATHER.PRECIP, shown below. There is one observation for each rainy day. The variable DATE stores the date of the rainfall. DATE is shown in the SAS DATE7. format. The variable PRECIP holds the total rainfall in inches. You need to produce an annual rainfall report showing total precipitation for each month.

SAS data set WEATHER.PRECIP

OBS	DATE	PRECIP
1	01JAN93	0.80
2	03JAN93	0.28
3	08JAN93	0.99
4	10JAN93	0.11
5	11JAN93	0.76
6	14JAN93	0.24
7	23JAN93	0.93
8	24JAN93	0.40
9	27JAN93	0.35
10	29JAN93	0.56
11	07FEB93	0.86
12	08FEB93	0.72
13	12FEB93	0.82
14	18FEB93	0.76
15	21FEB93	0.34
16	02MAR93	1.23
17	04MAR93	0.53
18	15MAR93	0.68
19	19MAR93	0.89
20	20MAR93	0.22
.		
.		
.		
66	07OCT93	0.09
67	08OCT93	0.31
68	12OCT93	0.57
69	16OCT93	0.78
70	20OCT93	0.28
71	31OCT93	0.70
72	02NOV93	0.56
73	03NOV93	0.53
74	08NOV93	0.42
75	13NOV93	0.20
76	15NOV93	0.68
77	25NOV93	0.23
78	28NOV93	0.50
79	29NOV93	0.91
80	30NOV93	0.74
81	03DEC93	0.90
82	17DEC93	0.24
83	18DEC93	0.78
84	30DEC93	0.30

MEANS Procedure and FORMAT Statement

To produce a monthly rainfall report, you need to group the daily rainfall data by month. In WEATHER.PRECIP, the variable DATE holds a SAS date value, that is, the number of days between January 1, 1960 and the given date value. Remember that the values of DATE, in the listing of WEATHER.PRECIP on the previous page, are displayed in the DATE7. format, which shows the number of days in the *ddmmyy* layout. The stored value is still just a number. There is no way to directly extract the name of the month from this numeric value.

The FORMAT statement provides the solution to the day-to-month grouping problem. You can use the FORMAT statement with many SAS procedures. When you do, the procedure will process the formatted values of a variable rather than the actual values. There are many SAS date formats you can assign with the FORMAT statement; one of them is MONYY5. The MONYY5. format converts a date from the number of days since January 1, 1960 to a string of characters made up of a three-character month abbreviation (JAN, FEB, MAR, etc.) and a two-character year (92, 93, 94, etc.). For example, the date August 4, 1993 would be represented as "AUG93". The "5." in MONYY5. refers to the total length of five characters. The length component of all SAS formats is followed by a period.

You can use PROC MEANS with the MONYY5. format to group the daily precipitation by month.

In this example, PROC MEANS is used to find the total rainfall for each month. **The MEANS procedure can be used to find many other statistics** such as mean (average), minimum, maximum, standard deviation, and others. PROC MEANS can also produce an output SAS data set instead of printed output. See Chapters 6 and 7 in this book and Chapter 21, "The MEANS Procedure," *SAS Procedures Guide, Version 6, Third Edition*.

Program 8

```
1  PROC MEANS DATA=WEATHER.PRECIP SUM MAXDEC=2;
2     TITLE 'Monthly Precipitation';
3     FORMAT DATE MONYY5.;
4     CLASS DATE;
5     VAR PRECIP;
6  RUN;
```

Program 8 Notes

Line

1. The PROC statement starts the MEANS procedure and names WEATHER.PRECIP as the input SAS data set. "SUM" tells PROC MEANS to generate the sum statistic, that is, to calculate totals. MAXDEC=2 limits output to two decimal places.

2. The TITLE statement applies to the PROC MEANS output. The quoted text will be displayed at the top of each page of output.

3. The FORMAT statement tells PROC MEANS to use the formatted values of the CLASS statement variable, not the actual values. A FORMAT statement contains one or more sets of variable names followed by a SAS format specification:

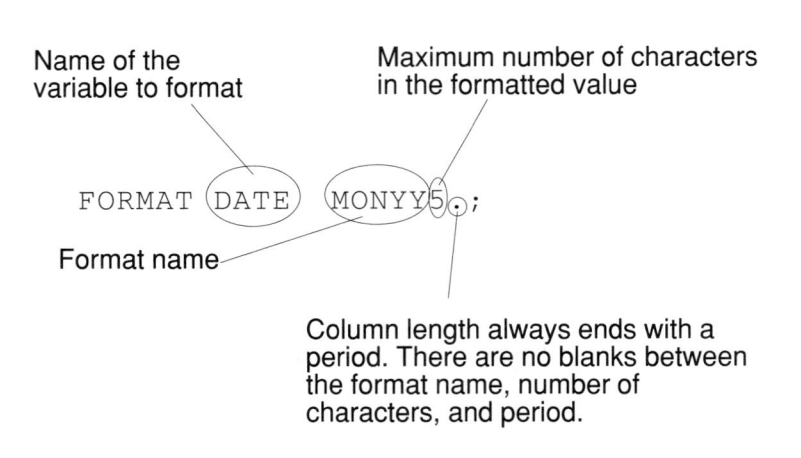

    ```
    FORMAT DATE MONYY5.;
    ```

    - DATE — Name of the variable to format
    - MONYY — Format name
    - 5 — Maximum number of characters in the formatted value
    - . — Column length always ends with a period. There are no blanks between the format name, number of characters, and period.

Program 8 Notes

Line

4 The CLASS statement tells PROC MEANS to group the data by the DATE variable. Because a FORMAT statement is also present, PROC MEANS will use the formatted values of DATE.

Without the FORMAT statement, the MEANS procedure would create a category for each unique value of the DATE variable in WEATHER.PRECIP. Since there are 84 observations in the data set, each with a unique DATE value, there would be 84 data groups in the output report with 1 observation in each group. The FORMAT statement converts observations 1-10 to "JAN93", observations 11-15 to "FEB93", etc., creating twelve unique values, one for each month.

5 The VAR statement names the variable to analyze, in this case PRECIP.

6 The RUN statement ends and executes the MEANS procedure.

SAS date formats have two components: a name and a width. In the MONYY5. format, "MONYY" is the format name, and "5." is the width. Width refers to the maximum number of characters that will be displayed. Formats have default widths. The default for MONYY is 5, so "MONYY." and "MONYY5." produce identical results. You can control the action of some SAS formats by changing the width. See End Notes for an example. See Chapter 14, "SAS Formats" in *SAS Language Reference, Version 6, First Edition* for full documentation of SAS formats.

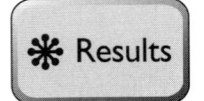

Results: Monthly Precipitation Report

The results of Program 8 appear in the monthly precipitation report below. Note the DATE groups. The N Obs column indicates the number of observations in each month. There were 10 rain days in January, 5 in February, etc. The Sum column is the total rainfall for each month. "Monthly Precipitation" comes from the TITLE statement (line 2) in Program 8. The "Analysis Variable : PRECIP" sub-title is added by PROC MEANS.

Monthly precipitation report

```
                  Monthly Precipitation

                 Analysis Variable : PRECIP


             DATE    N Obs           Sum
             ---------------------------------
             JAN93    10            5.42

             FEB93     5            3.50

             MAR93     8            5.14

             APR93     7            3.41

             MAY93     5            2.79

             JUN93     8            4.86

             JUL93     3            1.57

             AUG93     7            4.02

             SEP93    10            4.33

             OCT93     8            4.09

             NOV93     9            4.77

             DEC93     4            2.22
             ---------------------------------
```

Displaying Dates and Days of the Week

Other date formats. Several other SAS date formats are useful for grouping data gathered on a daily basis:

QTR*w*.	- quarter of the year, 1 to 4
MONTH*w*.	- month of the year, 1 to 12
MONNAME*w*.	- name of the month only, no year
YEAR*w*.	- year only
WEEKDAY*w*.	- day of the week, 1 to 7 (see next End Note)
WEEKDATE*w*.	- spelled-out date (see next End Note)

The WEEKDATE format. The WEEKDATE*w*. format is useful when you want to group data by the day of the week with the day name written out: "Sunday", "Monday", "Tuesday", etc. For example, you can display 8/9/93 as "Monday, August 9, 1993". You control how much of the date is displayed by altering the *w*. component of the format. If you specify the format WEEKDATE9., only the day name will be displayed or used by classifying PROCs like PROC MEANS. A width of 9 is just enough for "Wednesday", the longest weekday name. If you want to find the total rainfall for each of the seven weekdays, you can do it easily with the following MEANS procedure step:

```
PROC MEANS DATA=WEATHER.PRECIP SUM MAXDEC=2;
   TITLE 'Day of the Week Precipitation';
   FORMAT DATE WEEKDATE9.;
   CLASS DATE;
   VAR PRECIP;
RUN;
```

The following ouput is generated, with rainfall summed for each weekday:

```
Day of the Week Precipitation
Analysis Variable : PRECIP

      DATE   N Obs          Sum
-----------------------------------
  Tuesday      16         10.25
  Thursday     14          6.27
  Friday       11          6.95
  Monday       11          5.47
  Wednesday    11          6.36
  Sunday        9          4.38
  Saturday     12          6.44
-----------------------------------
```

Note that the order of the days results from the order that data were encountered in the data set WEATHER.PRECIP. Ordering days from Sunday to Saturday requires a different strategy. You could use the WEEKDAY*w.* format, which generates an ordinal number from 1 to 7, representing Sunday through Saturday. If you also include the ORDER=FORMATTED option in the PROC MEANS statement, the output will be in ascending order of the formatted value of the CLASS statement variable:

```
PROC MEANS DATA=WEATHER.PRECIP SUM MAXDEC=2 ORDER=FORMATTED;
    TITLE 'Day of the Week Precipitation';
    FORMAT DATE WEEKDAY1.;
    CLASS DATE;
    VAR PRECIP;
RUN;
```

The PROC MEANS step above generates the following output:

```
        Day of the Week Precipitation
           Analysis Variable : PRECIP


        DATE   N Obs            Sum
        ------------------------------
          1      9              4.38
          2     11              5.47
          3     16             10.25
          4     11              6.36
          5     14              6.27
          6     11              6.95
          7     12              6.44
        ------------------------------
```

Grouping Data: Days Into Months

1. Make sure you know the name and have access to the SAS data set holding the data you want to examine. This may require a LIBNAME statement. (See Chapter 2 in this book for a LIBNAME example.)

2. Make sure you know the names of the SAS variables with which you need to work.

3. Use PROC MEANS or another SAS procedure that can use the formatted values of variables with a FORMAT statement.

4. In the FORMAT statement, name the classification variable followed by the SAS format you want the procedure to use. To group dates by month, use the MONYY5. or MONNAME*w.* format.

Date Formats, PROC MEANS

FORMAT statement
"FORMAT" in Chapter 9, "SAS Language Statements," *SAS Language: Reference, Version 6, First Edition*
list of date formats
"Categories of Formats" in Chapter 3, "Components of the SAS Language," *SAS Language: Reference*
PROC MEANS
Chapter 21, "The MEANS Procedure," *SAS Procedures Guide, Version 6, Third Edition*
SAS formats
"SAS Formats" in Chapter 3, "Components of the SAS Language," *SAS Language: Reference*

Chapter 9
Grouping Data: Create Your Own Groups

Learn How to...

- Create customized data groups
- Use custom groups in a report

Using These SAS System Features...

- FORMAT procedure
- FORMAT statement
- FREQ procedure

? Problem   Compare Math Scores

To compare scores on the 8th grade final math test from three schools, you need to separate the scores into groups: Below Average, Average, and Above Average. The raw test scores from each school are contained in SAS data set SCORES.MATH8, shown below. There is one observation for each student. The variable SCHOOL indicates which school the student is from, and SCORE is the student's test score.

The Average score range has been determined to be 75-85, Above Average 86-100, and Below Average 0-74. You need to create a report showing the number and percentage of students in each school (North, Central, and Riverside) that fall into each of the three score ranges.

SAS data set SCORES.MATH8

OBS	SCHOOL	SCORE
1	NORTH	95
2	NORTH	79
3	NORTH	53
4	NORTH	76
5	NORTH	67
6	NORTH	90
7	NORTH	86
8	NORTH	71
9	NORTH	75
10	NORTH	73
11	NORTH	70
12	NORTH	58
13	NORTH	96
14	CENTRAL	84
15	CENTRAL	58
16	CENTRAL	55
17	CENTRAL	55
18	CENTRAL	78
19	CENTRAL	76
20	CENTRAL	80
21	CENTRAL	66
22	CENTRAL	54
23	CENTRAL	76
24	CENTRAL	84

SAS data set SCORES.MATH8 (continued)

OBS	SCHOOL	SCORE
25	CENTRAL	90
26	CENTRAL	81
27	CENTRAL	99
28	CENTRAL	66
29	CENTRAL	69
30	CENTRAL	87
31	CENTRAL	53
32	CENTRAL	76
33	CENTRAL	67
34	CENTRAL	65
35	CENTRAL	85
36	CENTRAL	85
37	CENTRAL	51
38	RIVERSIDE	68
39	RIVERSIDE	78
40	RIVERSIDE	98
41	RIVERSIDE	72
42	RIVERSIDE	84
43	RIVERSIDE	62
44	RIVERSIDE	80
45	RIVERSIDE	99
46	RIVERSIDE	59
47	RIVERSIDE	76
48	RIVERSIDE	90
49	RIVERSIDE	74
50	RIVERSIDE	92
51	RIVERSIDE	85
52	RIVERSIDE	82
53	RIVERSIDE	62
54	RIVERSIDE	69
55	RIVERSIDE	83
56	RIVERSIDE	64
57	RIVERSIDE	82
58	RIVERSIDE	79

PROC FORMAT and PROC FREQ

For the test score report you need to group the data at two levels; first at the school level, then by score category: Below Average, Average, or Above Average. The SAS System provides two ways of grouping data. One is the BY statement (see Chapter 4). The BY statement groups data that are in sorted order. A second method relies on classification procedures such as PROC MEANS and PROC FREQ that group data by the variables listed in a CLASS or TABLE statement (see Chapters 5 through 8). These methods can be combined in one procedure step.

Program 9 combines a BY statement and a FORMAT statement in a PROC FREQ step. When a FORMAT statement is used with PROC FREQ, the formatted value of the TABLE variable is used instead of the actual value. To take advantage of this feature, you need to create a format for the three test score categories, then tell PROC FREQ to use those categories instead of the actual scores.

Program 9 is made up of three procedure steps:

1. A PROC FORMAT step creates the customized categories: Below Average, Average, and Above Average.

2. The SORT procedure ensures that SCORES.MATH8 is sorted by school name.

3. PROC FREQ with BY and FORMAT statements generates the report.

PROC FORMAT is a powerful tool for creating data groups. You can use these custom formats just like standard SAS System formats in PROC FREQ, PROC MEANS, and other procedures.

Program 9

```
1   PROC FORMAT;
2     VALUE SCRFMT
3          0-74    = 'Below Average'
4          75-85   = 'Average'
5          86-100  = 'Above Average';
6   RUN;
7
8   PROC SORT DATA=SCORES.MATH8;
9      BY SCHOOL;
10  RUN;
11
12  PROC FREQ DATA=SCORES.MATH8;
13     TITLE '8th Grade Math Test Results';
14     FORMAT SCORE SCRFMT.;
15     BY SCHOOL;
16     TABLE SCORE;
17  RUN;
```

Program 9 Notes

Line   **1**   The PROC statement starts the FORMAT procedure.

2   The VALUE statement begins on this line with the keyword "VALUE" and ends on line 5 with the semicolon. "SCRFMT" names the custom format you want to create. The format name is up to you but it must be a valid SAS name up to eight characters. Do not end the format name with a period.

3-5   These lines are part of the VALUE statement. Each line creates a category. Line 3 creates a category called Below Average and assigns all values from 0 to 74 to it. Line 4 assigns values from 75 to 85 to a category called Average, and line 5 assigns values from 86 to 100 to Above Average. You must enclose category names in quotes in the VALUE statement.

Program 9 Notes

Line **6** The RUN statement ends and executes PROC FORMAT.

8 The PROC statement starts the SORT procedure and names SCORES.MATH8 as the input SAS data set. No output data set is named, so the original SCORES.MATH8 will be replaced with the sorted version.

9 The BY statement tells the SORT procedure which variable to sort on.

10 The RUN statement ends and executes PROC SORT.

12 The PROC statement starts the FREQ procedure and names SCORES.MATH8 as the input SAS data set.

13 The TITLE statement applies to the PROC FREQ output.

14 The FORMAT statement tells PROC FREQ to use the formatted values of the variable SCORE, not the actual values. The FORMAT statement contains the variable name (SCORE) and the corresponding format (SCRFMT.). Note that the format name must end with a period when you use it, even though you do not include a period when you create it with the PROC FORMAT VALUE statement (line 2).

15 The BY statement tells PROC FREQ to process SCORES.MATH8 in groups according to the value of SCHOOL. When you use a BY statement with a procedure, the input data set must be in BY-variable order. The PROC SORT step (lines 8 through 10) ensures the data set is sorted properly. (See End Notes for an alternative method.)

16 The TABLE statement names the variable you want to analyze. PROC FREQ counts the number of unique values of the variable SCORE. Since a FORMAT statement is included in the step, PROC FREQ will count each unique **formatted** value of SCORE.

17 The RUN statement ends and executes the PROC FREQ step.

Closer Look — The SCRFMT. format is temporary. **When you leave the SAS System the format disappears.** You can make a format permanent by storing it in a SAS catalog. This allows you to reuse the format later or share it with others. See "PROC FORMAT Statement" in Chapter 18, "The FORMAT Procedure," *SAS Procedures Guide, Version 6, Third Edition* and "Release 6.06: SAS Catalogs" in Chapter 6, "SAS Files," *SAS Language: Reference, Version 6, First Edition.*

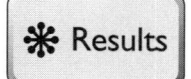

Test Score Report

✱ Results

Program 9 generates the test results report shown below. PROC FREQ generates a separate table for each BY group (SCHOOL). The TITLE statement text appears above each of these tables. Statistics represent the formatted values of SCORE. Within each BY group, `Frequency` is the count of students in each of three score categories, and `Percent` is the percent in each category. You can see that at the Riverside school, 42.9% of the students scored in the Average range. The cumulative statistics represent the current value of SCORE plus all preceding values.

Test score report

```
                         8th Grade Math Test Results

--------------------------- SCHOOL=CENTRAL ---------------------------

                                              Cumulative    Cumulative
            SCORE    Frequency    Percent     Frequency      Percent
            ---------------------------------------------------------
            Below Average     11       45.8          11         45.8
            Average           10       41.7          21         87.5
            Above Average      3       12.5          24        100.0


                         8th Grade Math Test Results

---------------------------- SCHOOL=NORTH ----------------------------

                                              Cumulative    Cumulative
            SCORE    Frequency    Percent     Frequency      Percent
            ---------------------------------------------------------
            Below Average      6       46.2           6         46.2
            Average            3       23.1           9         69.2
            Above Average      4       30.8          13        100.0


                         8th Grade Math Test Results

-------------------------- SCHOOL=RIVERSIDE --------------------------

                                              Cumulative    Cumulative
            SCORE    Frequency    Percent     Frequency      Percent
            ---------------------------------------------------------
            Below Average      8       38.1           8         38.1
            Average            9       42.9          17         81.0
            Above Average      4       19.0          21        100.0
```

Using SAS Formats and PROC FORMAT

Character formats. The format you created in Program 9 (SCRFMT.) is a numeric format because it translates numeric values (test scores) into character strings ("Below Average", etc.). You can also create character formats that translate character values. For example, you could create a format that translates "Y" to "Yes" and "N" to "No". Like all character formats, the name you choose for a character format must begin with a "$".

More on PROC FORMAT. There is much more to PROC FORMAT than this chapter demonstrates. For example, you can use the procedure to generate picture formats that add dashes or parentheses to telephone or Social Security numbers. PROC FORMAT can also create *informats* that translate values input from an external file or other source. See Chapter 18, "The Format Procedure" in the *SAS Procedures Guide, Version 6, Third Edition* for full documentation.

Grouped but not sorted. Normally when you use a BY statement with a SAS procedure, the input data set must be in BY-variable order. If you look at the SAS data set SCORES.MATH8 on the previous pages, you can see that the data set is grouped by school name but not sorted by the variable SCHOOL. All the NORTH scores appear together, but before the CENTRAL scores. If you know your data are properly grouped, you can avoid sorting by using the NOTSORTED option with the BY statement. For example, an alternative version of Program 9 without PROC SORT follows:

```
PROC FORMAT;
   VALUE SCRFMT
      0-74   = 'Below Average'
      75-85  = 'Average'
      86-100 = 'Above Average';
RUN;

PROC FREQ DATA=SCORES.MATH8;
   TITLE '8th Grade Math Test Results';
   FORMAT SCORE SCRFMT.;
   BY SCHOOL NOTSORTED;
   TABLE SCORE;
RUN;
```

The resulting output would show the North school results first, then Central and Riverside. Results would be in the same order as the groups in the SCORES.MATH8 data set. The PROC SORT step ensures that the input data set will be compatible with the BY statement, but sorting consumes system resources and processing time. There is no need to use a PROC SORT step if you know the data set is already sorted or grouped.

98 Part 2 - Working with Your Data

📝 **Lookup tables.** If you examine the PROC FORMAT step in Program 9, you can see that the VALUE statement information acts like an input/output lookup table with three rows and two columns:

Input Value	Output Value
0 to 74	"Below Average"
75 to 85	"Average"
86 to 100	"Above Average"

While there are some practical limits, the SAS System itself imposes no limit on the number of entries you can have in a VALUE statement. Many programmers use this capability to create large lookup tables. For example, you can generate a custom format that uses customer numbers on the input side and the full customer name on the output side. Such a table can have hundreds or thousands of entries:

```
VALUE CUSTNAME
      101 = "Pete's Auto Body"
      102 = "Isamu Parts and Repair"
      103 = "Keystone Automotive"
                  .
                  .
                  .
      800 = "Diversified Products Inc."
      801 = "Sunset Ford and Mazda"
      803 = "Riverside Auto Repair";
```

There are other techniques for implementing lookup tables in the SAS System. The primary advantage of using PROC FORMAT is performance. Pulling a value from a custom format is much faster than alternatives such as searching for the value in a SAS data set. For examples of PROC FORMAT lookup tables see "Example 1" in Chapter 18, "The Format Procedure," *SAS Procedures Guide*, and "Using Formatted Data Values as a Lookup File" in Chapter 10, "Performing a Table Lookup," *SAS Language and Procedures: Usage 2, Version 6, First Edition*.

Grouping Data: Create Your Own Groups

Quick Summary

1. Make sure you know the name and have access to the SAS data set holding the data you want to examine. This may require a LIBNAME statement. (See Chapter 2 in this book for a LIBNAME example.)

2. Make sure you know the names of the SAS variables with which you need to work.

3. Use PROC FORMAT to create a table of input value ranges and output values. Name the custom format and assign input/output pairs in a VALUE statement. A single VALUE statement can have many input/output pairs.

4. Use your custom format in SAS procedures that support formatted values such as PROC FREQ and PROC MEANS.

Formats and the FORMAT Procedure

More Information

FORMAT statement
"FORMAT" in Chapter 9, "SAS Language Statements," *SAS Language: Reference, Version 6, First Edition*

FREQ procedure
Chapter 20, "The FREQ Procedure," *SAS Procedures Guide, Version 6, Third Edition*

PROC FORMAT
"Categorizing Variable Values with Customized Formats" in Chapter 8, "Classifying Variables into Categories," *SAS Language and Procedures: Usage 2, Version 6, First Edition;*
Chapter 18, "The Format Procedure," *SAS Procedures Guide*

Chapter 10
Combining SAS® Data Sets

Learn How to...

- Combine data from two or more data sets
- Read observations from multiple data sets
- Check the origin of an observation

Using These SAS System Features...

- SAS DATA step
- SET statement
- IN= data set option
- SELECT statement

? Problem — Combine Experimental Results

To test the effectiveness of an antibiotic, a bacteria growth study was broken down into three separate experiments conducted under differing conditions. In each experiment a bacteria count was taken every hour for ten hours. The resulting data were entered into a SAS data set with two variables: HOUR and COUNT. Three separate SAS data sets were created, one for each experiment. The data sets are RESULTS.EX91102, RESULTS.EX91103, and RESULTS.EX91104, which appear below and on the next page.

To analyze the data, you need to consolidate the information from the experiments into a single SAS data set. Because each experiment was conducted under different conditions, the combined data set must contain a variable that identifies the source of each observation.

SAS data set RESULTS.EX91102

OBS	COUNT	HOUR
1	118	1
2	165	2
3	234	3
4	310	4
5	347	5
6	355	6
7	404	7
8	479	8
9	519	9
10	536	10

SAS data set RESULTS.EX91103

OBS	COUNT	HOUR
1	220	1
2	287	2
3	334	3
4	387	4
5	415	5
6	486	6
7	494	7
8	584	8
9	644	9
10	706	10

SAS data set RESULTS.EX91104

OBS	COUNT	HOUR
1	88	1
2	135	2
3	234	3
4	271	4
5	293	5
6	373	6
7	388	7
8	487	8
9	560	9
10	586	10

Solution: Data Step, SET Statement, IN= Data Set Option

This problem involves concatenating SAS data sets, that is, putting them together end-to-end. It's easy to concatenate data sets with a DATA step by simply naming each data set in a SET statement. The SET statement reads observations from a SAS data set. If you list more than one data set, it will read all observations in the first data set then continue to the next. The DATA step writes each observation read by the SET statement to the output SAS data set.

The experimental results problem is complicated by the fact that you need to retain the identity of each observation. If the three data sets were simply concatenated by reading one after the other, there would be no way to distinguish an observation from the RESULTS.EX91102 data set from observations originating in one of the other data sets. You need to add a third variable to identify the origin of each observation. To do this, you can use the SELECT statement and IN= data set option. When reading from RESULTS.EX91102, the text "FROM102" is assigned to this third variable; when reading from RESULTS.EX91103, "FROM103" is assigned, etc. The actual values assigned are not important as long as you use three different values.

Program 10

```
1   DATA RESULTS.ALLEXPER;
2
3      SET RESULTS.EX91102(IN=IN102)
4          RESULTS.EX91103(IN=IN103)
5          RESULTS.EX91104(IN=IN104);
6
7      SELECT;
8         WHEN (IN102 = 1) EXPER = 'FROM102';
9         WHEN (IN103 = 1) EXPER = 'FROM103';
10        WHEN (IN104 = 1) EXPER = 'FROM104';
11     END;
12
13  RUN;
14
15  PROC PRINT DATA=RESULTS.ALLEXPER;
16     TITLE 'Bacteria Growth';
17     TITLE2 'Experimental Results';
18  RUN;
```

Closer Look

IN= is one of several data set options. SAS data set options are named inside parentheses immediately following the SAS data set name. These options can be very useful when you are dealing with multiple data sets in a single DATA or PROC step because they apply only to the data set to which they are associated, rather than all data sets used in the step. Some data set options are:

DROP=   - drops variables from the associated data set.
KEEP=   - keeps listed variables in the associated data set; all others are dropped.
OBS=   - ends processing after the specified observation number has been read.
RENAME= - renames variables in the associated data set.

See Chapter 15, "SAS Data Set Options," in *SAS Language: Reference, Version 6, First Edition* for details.

Program 10 Notes

Line

1 The DATA statement starts the DATA step and names the output SAS data set, RESULTS.ALLEXPER. This will be the new, combined data set.

3 The SET statement begins on line 3 and ends with the semicolon on line 5. The SET statement reads observations from the three SAS data sets. Observations are read one at a time beginning with the first data set listed. All observations in RESULTS.EX91102 are read before reading RESULTS.EX91103. All observations in RESULTS.EX91103 are read before reading RESULTS.EX91104.

The IN= option tells the SAS System to assign a value of 1 to the variable IN102 when the SET statement reads an observation from RESULTS.EX91102; otherwise assign 0. "IN102" is an arbitrary variable name. You could use any valid SAS name.

4 Here, the IN= option tells the SAS System to assign a value of 1 to the variable IN103 when the SET statement reads an observation from RESULTS.EX91103; otherwise assign 0. Like "IN102" above, "IN103" is an arbitrary variable name.

5 The IN= option tells the SAS System to assign a value of 1 to IN104 when the SET statement reads an observation from RESULTS.EX91104; otherwise assign 0. The semicolon ends the SET statement.

7 The SELECT statement opens a series of statements called a SELECT group. The group is terminated by the END statement on line 11. A SELECT group represents one or more alternatives. Here, three alternatives are offered on lines 8, 9, and 10.

8 This WHEN statement represents the first alternative in the SELECT group. It tests the value of IN102. When IN102=1 the variable EXPER is assigned the value "FROM102". If IN102≠1 the next alternative in the SELECT group is checked.

9-10 These WHEN statements are the second and third alternatives in the SELECT group. When the variable IN103 has a value of 1, then EXPER is set to "FROM103"; if not, then the variable IN104 is checked. The values of the variables IN102, IN103, and IN104 are determined by the IN= data set options in the SET statement (lines 3–5).

The values "FROM102", "FROM103", and "FROM104" assigned to the variable EXPER are arbitrary. You could use "A", "B", and "C" or numeric values 1, 2, and 3, for example.

Program 10 Notes

Line **11**   The END statement ends the SELECT group.

13   The RUN statement ends and executes the DATA step.

15   The PROC statement starts the PRINT procedure and names RESULTS.ALLEXPER, created in the previous DATA step, as input.

16   The TITLE statement applies to the PROC PRINT output. The quoted text will be displayed at the top of each page.

17   The TITLE2 statement applies to the PROC PRINT output. The quoted text will be displayed as the second line at the top of each page.

18   The RUN statement ends and executes the PRINT procedure.

Closer Look

There is another way to use the SELECT group. You can add an expression to the SELECT statement then test for alternative values of the expression in WHEN statements:

```
SELECT (X);
      WHEN (1) SOMEVAR = 'X was 1';
      WHEN (2) SOMEVAR = 'X was 2';
      WHEN (3) SOMEVAR = 'X was 3';
      OTHERWISE SOMEVAR = 'X was ?';
END;
```

This SELECT group examines the value of the variable X named in parenthesis in the SELECT statement. If X=1 then SOMEVAR is set to "X was 1". If X=2 then SOMEVAR is set to "X was 2"; if X=3 then SOMEVAR is set to "X was 3". In all other cases SOMEVAR is set to "X was ?". Note that in any SELECT group, one of the conditions in the group must be true. If you left out the OTHERWISE statement in the group above and the value of X was 4, the system would issue an error.

✳ Results

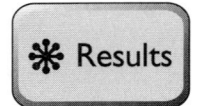

SAS Data Set RESULTS.ALLEXPER

Observations from the three experimental results data sets are combined in the data set RESULTS.ALLEXPER, which appears below in the PROC PRINT output. Note the variables COUNT and HOUR are present, plus the variable EXPER created in the DATA step. The values of EXPER were assigned in the DATA step SELECT group (lines 7-11). EXPER allows you to identify the origin of each observation.

SAS data set RESULTS.ALLEXPER

```
                 Bacteria Growth
               Experimental Results

        OBS     COUNT     HOUR      EXPER

          1      118        1       FROM102
          2      165        2       FROM102
          3      234        3       FROM102
          4      310        4       FROM102
          5      347        5       FROM102
          6      355        6       FROM102
          7      404        7       FROM102
          8      479        8       FROM102
          9      519        9       FROM102
         10      536       10       FROM102
         11      220        1       FROM103
         12      287        2       FROM103
         13      334        3       FROM103
         14      387        4       FROM103
         15      415        5       FROM103
         16      486        6       FROM103
         17      494        7       FROM103
         18      584        8       FROM103
         19      644        9       FROM103
         20      706       10       FROM103
         21       88        1       FROM104
         22      135        2       FROM104
         23      234        3       FROM104
         24      271        4       FROM104
         25      293        5       FROM104
         26      373        6       FROM104
         27      388        7       FROM104
         28      487        8       FROM104
         29      560        9       FROM104
         30      586       10       FROM104
```

SET Statement, PROC APPEND

SET statement. It is important to understand the SET statement if you plan on using the DATA step to process SAS data sets. The SET statement reads observations from the SAS data set or sets listed in the statement. SET reads only from SAS data sets, not external files (see Chapter 2 on external files). In Program 10, the SET statement is used to read three SAS data sets, one after the other. It is more commonly used to read a single data set so some operation can be performed on each observation in that data set. For example, if you want to add 12 to the HOUR variable in RESULTS.ALLEXPER, you can use the following DATA step to read each observation and perform the addition:

```
DATA RESULTS.ALLEXPER;
    SET RESULTS.ALLEXPER;
    HOUR=HOUR+12;
RUN;
```

First, the SET statement reads an observation from RESULTS.ALLEXPER. Next, 12 is added to the value of HOUR in that observation. The altered observation is automatically written to a new version of RESULTS.ALLEXPER. The DATA step loops to the top and the SET statement reads the next available observation. This sequence continues until no more observations are available to the SET statement. After processing the last observation, the DATA step ends.

PROC APPEND. You can also concatenate SAS data sets with the APPEND procedure. PROC APPEND works with two data sets at a time: a base data set and a new data set that is added to the end of the base data set. If you wanted to concatenate RESULTS.EX91103 to the end of RESULTS.EX91102, you could use the following step:

```
PROC APPEND BASE=RESULTS.EX91102
    NEW=RESULTS.EX91103;
RUN;
```

After you run this step, RESULTS.EX91102 contains all the observations from the original RESULTS.EX91102 plus all the observations from RESULTS.EX91103. RESULTS.EX91103 remains unchanged. PROC APPEND cannot add a variable that tags the observations by origin such as in Program 10. See Chapter 5, "The APPEND Procedure" in *SAS Procedures Guide, Version 6, Third Edition* for more information.

Combining SAS Data Sets

1. Make sure you know the names and have access to the SAS data sets you want to combine. This may require a LIBNAME statement. (See Chapter 2 in this book for a LIBNAME example.)

2. Use a DATA step. Name the new, combined SAS data set in the DATA statement.

3. Use a SET statement listing the data sets you want to concatenate.

4. If you need to identify the origin of the observations, use the IN= data set option. Use the SELECT statement to test the IN= variables and assign values to a tag variable that identifies the source of each observation.

Combining Data Sets with the DATA Step

combining data sets overview
Chapter 15, "Concatenating SAS Data Sets," *SAS Language and Procedures: Usage, Version 6, First Edition*
"Combining SAS Data Sets" in Chapter 4, "Rules of the SAS Language," *SAS Language: Reference, Version 6, First Edition*

DATA step overview
Chapter 2, "The DATA Step," *SAS Language: Reference*

IN= data set option
"Using the IN= Data Set Option" in Chapter 19, "Manipulating SAS Data Sets," *SAS Language and Procedures: Usage*
"IN=" in Chapter 15, "SAS Data Set Options," *SAS Language: Reference*

SELECT statement
"SELECT" in Chapter 9, "SAS Language Statements," *SAS Language: Reference*

SET statement
"SET" in Chapter 9, "SAS Language Statements," *SAS Language: Reference*

Chapter 11
Selecting and Changing Data

Learn How to...

- Read a SAS data set
- Select specific observations
- Change the value of a variable

Using These SAS System Features...

- DATA step
- SET statement
- IF statement
- Assignment statement

? Problem: Fix the Shipping Department Log

The shipping department of a small business logs each product shipment in the SAS data set SDEPT.SHIPLOG, shown below as PROC PRINT output. A mistake was discovered for order number "54895" (observation 6). The date entered was one week too early. You need to write a program to find this entry and correct the shipping date.

In the SDEPT.SHIPLOG data set, the variable SHIPDATE holds the shipping date as a SAS date value and has been assigned the DATE7. format. CUSTNUM holds the customer number and ORDERNUM holds the order number. CUSTNUM and ORDERNUM are character variables. AMOUNT holds the dollar value of the order.

SAS data set SDEPT.SHIPLOG

OBS	SHIPDATE	CUSTNUM	ORDERNUM	AMOUNT
1	14JUL93	81883	54880	1684.49
2	01AUG93	22090	54881	4871.26
3	18JUL93	81883	54882	2226.68
4	27JUL93	76341	54884	981.37
5	18JUL93	81883	54894	1266.57
6	18JUL93	81883	54895	8598.95
7	15JUL93	76341	54902	5138.64
8	19JUL93	81883	54903	602.92
9	17JUL93	97123	54905	3489.04
10	09JUL93	76341	54906	4292.98
11	03AUG93	76341	54908	6281.53
12	27JUL93	97123	54909	6669.70
13	04AUG93	97123	54911	9453.24
14	15JUL93	76341	54916	9598.68
15	05AUG93	76341	54919	2041.23
16	19JUL93	22090	54920	2012.79
17	27JUL93	22090	54926	4806.97
18	30JUL93	76341	54927	2870.91
19	01AUG93	76341	54929	5942.01
20	08JUL93	22090	54930	4876.46
21	11JUL93	76341	54932	5362.34
22	13JUL93	81883	54933	4492.34
23	17JUL93	97123	54935	8005.08
24	02AUG93	81883	54936	7086.81
25	03AUG93	97123	54937	1667.25
26	31JUL93	97123	54938	3915.88

Solution: DATA Step, IF and Assignment Statements

To fix the shipment data set you need to find the observation in error then add 7 to the value of SHIPDATE in that observation. Why 7? SHIPDATE is a SAS date value; this means it is the number of days since January 1, 1960. Adding seven days moves the date one week forward.

Program 11 uses a SAS DATA step with a SET statement to read each observation in SDEPT.SHIPLOG. The value of ORDERNUM is tested with an IF statement. When ORDERNUM is "54895", 7 is added to SHIPDATE. Note that when you use a SAS data set as both input (named in the SET statement) and output (named in the DATA statement) an error or computer problem could result in a loss of the original data. See End Notes in this chapter for more on data set replacement and protection.

Program 11

```
1  DATA SDEPT.SHIPLOG;
2     SET SDEPT.SHIPLOG;
3     IF ORDERNUM='54895' THEN SHIPDATE=SHIPDATE+7;
4  RUN;
5
6  PROC PRINT DATA=SDEPT.SHIPLOG;
7     TITLE 'Corrected Shipped Orders';
8  RUN;
```

Closer Look

In the data set SDEPT.SHIPLOG, **ORDERNUM is a character variable, not numeric.** ORDERNUM is checked for the string of characters: "5", "4", "8", "9", and "5", not the number 54,895. When you create a data set that contains identifiers like employee or customer numbers, you can use character variables if you do not intend to perform a mathematical operation on the variable. You should also consider how the values will be used. If the last two digits of a customer number indicate which state the customer is located in, it will be easier to pull out this information if the value is character. Several features of the SAS DATA step language make it easy to convert values from character to numeric and vice versa, so you can usually change data types if necessary. See "Variables in Assignment Statements" in Chapter 3, "Components of the SAS Language," *SAS Language: Reference, Version 6, First Edition.*

Program 11 Notes

Line 1 The DATA statement starts the DATA step and names SDEPT.SHIPLOG as the output SAS data set. SDEPT.SHIPLOG is also the input data set. The original SDEPT.SHIPLOG will be replaced by the new version with the corrected date. (See End Notes.)

2 The SET statement reads SDEPT.SHIPLOG one observation at a time in first-to-last sequence.

3 The IF statement tests the value of ORDERNUM in each observation read by the preceding SET statement. If ORDERNUM is "54895", then the assignment statement SHIPDATE=SHIPDATE+7 is executed. The assignment statement adds seven days to the SAS date, moving the date up by a week, as required to fix the error. When used in SAS program statements, character literals must be enclosed in quotes. The quotes tell the system to treat the value as a string of characters, not a variable name or number.

4 The RUN statement ends and executes the DATA step.

6 The PROC statement begins the PROC PRINT step and names SDEPT.SHIPLOG as the input data set.

7 The TITLE statement applies to the PROC PRINT output. The quoted text will be displayed at the top of each page.

8 The RUN statement ends and executes the PRINT procedure.

Closer Look

Before the DATA step runs, **the SAS System adds implicit instructions to your program.** You can think of the DATA step in Program 11 as made up of the following instructions:

1. SET SDEPT.SHIPLOG (reads an observation).
2. *If no more data then end the DATA step.*
3. IF ORDERNUM = '54895' THEN SHIPDATE = SHIPDATE + 7.
4. *Write an observation to data set SDEPT.SHIPLOG.*
5. *Go to the top of the DATA step, instruction 1.*

Instructions 1 and 3 are explicit statements in the Program 11 DATA step. Instructions 2, 4, and 5 are implicit instructions added by the SAS System. There are other implicit instructions not shown here. The addition of implicit instructions is one reason the SAS DATA step language is so efficient. It's also the reason some results can seem a bit magical. See "DATA Step Processing" in Chapter 2, "The DATA Step," *SAS Language: Reference* for a discussion of DATA step processing and flow.

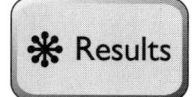

✱ Results Corrected Shipments Log

The corrected shipping log appears below in the output from the PROC PRINT step (lines 6–8). The shipping date for order number "54895" (observation 6) in the original data set has been changed to July 25, 1993 from July 18, 1993. All other observations are unchanged. Only observation 6 satisfied the IF condition on line 3.

Corrected SDEPT.SHIPLOG data set

```
                    Corrected Shipped Orders

       OBS    SHIPDATE    CUSTNUM    ORDERNUM      AMOUNT

        1     14JUL93      81883      54880       1684.49
        2     01AUG93      22090      54881       4871.26
        3     18JUL93      81883      54882       2226.68
        4     27JUL93      76341      54884        981.37
        5     18JUL93      81883      54894       1266.57
        6     25JUL93      81883      54895       8598.95
        7     15JUL93      76341      54902       5138.64
        8     19JUL93      81883      54903        602.92
        9     17JUL93      97123      54905       3489.04
       10     09JUL93      76341      54906       4292.98
       11     03AUG93      76341      54908       6281.53
       12     27JUL93      97123      54909       6669.70
       13     04AUG93      97123      54911       9453.24
       14     15JUL93      76341      54916       9598.68
       15     05AUG93      76341      54919       2041.23
       16     19JUL93      22090      54920       2012.79
       17     27JUL93      22090      54926       4806.97
       18     30JUL93      76341      54927       2870.91
       19     01AUG93      76341      54929       5942.01
       20     08JUL93      22090      54930       4876.46
       21     11JUL93      76341      54932       5362.34
       22     13JUL93      81883      54933       4492.34
       23     17JUL93      97123      54935       8005.08
       24     02AUG93      81883      54936       7086.81
       25     03AUG93      97123      54937       1667.25
       26     31JUL93      97123      54938       3915.88
```

Changing SAS Data Sets, IF Statements

Replacing SAS data sets with the DATA step. When you name a SAS data set as both input (in the SET statement) and output (in the DATA statement), the original version is replaced by a new version. As the DATA step runs, each observation is read by the SET statement, and any subsequent DATA step statements are executed. The observation is written to the new verison of the data set at the end of the DATA step. (See Closer Look after Program Notes.)

When you use the DATA step with a SET statement to replace a data set, the SAS System does not update observations in place while the DATA step is running. Instead, a temporary data set is created to hold the new observations. When the DATA step ends, the original data set is deleted and the temporary data set is given the original name. One consequence of this process is that, while the DATA step is running, you need enough free disk space to hold both the temporary and original data sets. In fact, this is true any time you replace a SAS data set, whether in a DATA step or PROC step, such as PROC SORT. When dealing with large SAS data sets, this temporary disk space requirement can be significant. If in doubt, check the size of the data sets you're dealing with against available space. You can use the CONTENTS procedure to check data set size. (See Chapter 12, "The CONTENTS Procedure" in *SAS Procedures Guide, Version 6, Third Edition.*) PROC CONTENTS will tell you the number of observations and the length of each. You can multiply these to get an approximation of space requirements. You can also use host-system commands to check file size if you know the system filename of the SAS data sets you're using.

Preserving input SAS data sets. Sometimes you want to change information in a SAS data set yet preserve the original data. To do this, simply name a new data set in the DATA statement as follows:

```
DATA SDEPT.FIXEDLOG;
   SET SDEPT.SHIPLOG;
   IF ORDERNUM='54895' THEN SHIPDATE=SHIPDATE+7;
RUN;
```

The program above reads SDEPT.SHIPLOG and creates a new SAS data set, SDEPT.FIXEDLOG. The data set named in the DATA statement is always the output data set. At most SAS software installations, if the output data set already exists it is overwritten without warning. As a further precaution you can use the SAS system option NOREPLACE to prevent the replacement of permanent SAS data sets. See "REPLACE" in Chapter 16, "SAS System Options," *SAS Language: Reference.* Always have a backup copy of any data you value.

IF statements. IF-THEN/ELSE statements are a basic component of most computer languages. The IF statement used in the DATA step works like IF statements in other procedural programming languages such as BASIC or COBOL. In the SAS System, ELSE is not part of the IF statement. It is a separate statement that begins with the keyword ELSE and ends with a semicolon:

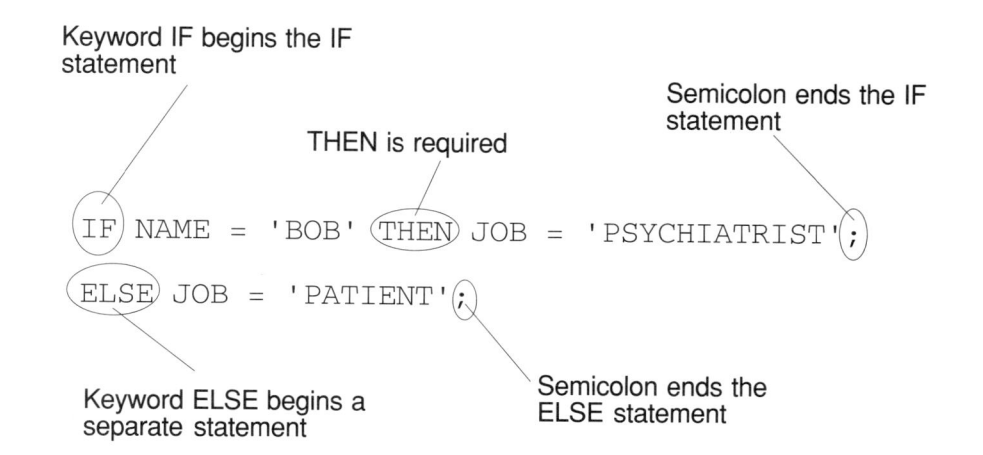

```
IF NAME = 'BOB' THEN JOB = 'PSYCHIATRIST';
ELSE JOB = 'PATIENT';
```

- Keyword IF begins the IF statement
- THEN is required
- Semicolon ends the IF statement
- Keyword ELSE begins a separate statement
- Semicolon ends the ELSE statement

You can execute a group of statements when the IF condition is true with a DO group:

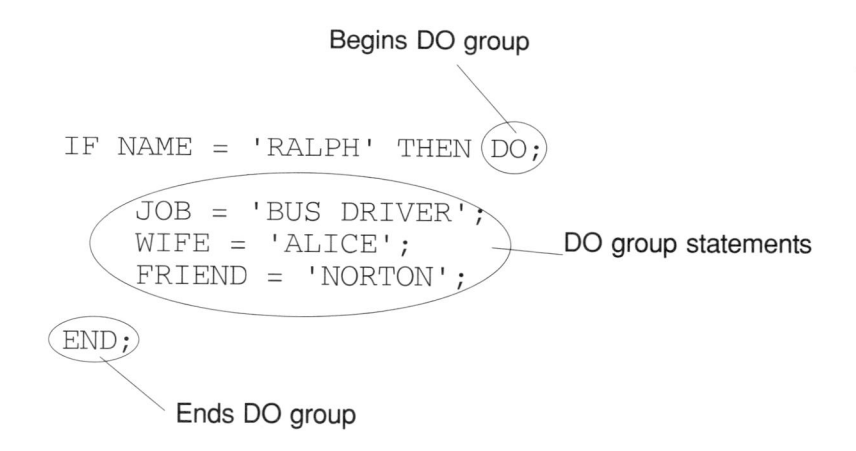

```
IF NAME = 'RALPH' THEN DO;
    JOB = 'BUS DRIVER';
    WIFE = 'ALICE';
    FRIEND = 'NORTON';
END;
```

- Begins DO group
- DO group statements
- Ends DO group

DO groups must be terminated with an END statement.

You can use the SELECT statement as an alternative to IF-THEN/ELSE. See Chapter 10 in this book for an example.

Selecting and Changing Data

1. Make sure you know the names and have access to the SAS data sets you want to combine. This may require a LIBNAME statement. (See Chapter 2 in this book for a LIBNAME example.)

2. Use a DATA step with the SET statement to read observations in the data set.

3. Select data to change with the IF statement.

4. Use an assignment statement to change values when the IF condition is true. Character value literals must be enclosed in quotes.

Basic DATA Step Programming Statements

assignment statements
"Adding Information to Observations with a DATA Step" in Chapter 5, "Understanding DATA Step Processing," *SAS Language and Procedures: Usage, Version 6, First Edition;*
"Assignment" in Chapter 9, "SAS Language Statements," *SAS Language: Reference, Version 6, First Edition*
DATA statement
"DATA" in Chapter 9, "SAS Language Statements," *SAS Language: Reference*
DATA step
Chapter 2, "The DATA Step," *SAS Language: Reference*
IF-THEN/ELSE statements
"How the Selection Process Works" in Chapter 8, "Acting on Selected Observations," *SAS Language and Procedures: Usage;*
"Performing More than One Action in One IF-THEN Statement" in Chapter 12, "Finding Shortcuts in Programming," *SAS Language and Procedures: Usage;*
"IF-THEN/ELSE" in Chapter 9, "SAS Language Statements," *SAS Language: Reference*

118 Part 2 - Working with Your Data

Chapter 12
Finding Unique Values

Learn How to...

- Find one-and-only-one values in a data set
- Create a list of unique values

Using These SAS System Features...

- DATA step
- SET and BY statements
- FIRST. and LAST. temporary variables
- Subsetting IF statement

Problem: Generate a Low Inventory Report

Super Sleuth Books buys used and out-of-print mystery books for resale. Each book is entered in an inventory data set called SLEUTH.BOOKS, which appears below and on the next page. To guide the store's book buyer, you need to create a list of authors for which there is only one book remaining in inventory.

SAS data set SLEUTH.BOOKS

```
                        Book List

     OBS    AUTHOR                    TITLE

      1     ALLBEURY, TED             THE LANTERN NETWORK
      2     ALLBEURY, TED             THE LANTERN NETWORK
      3     BARNES, LINDA             THE SNAKE TATTOO
      4     BOYER, RICK               THE DAISY DUCKS
      5     BOYER, RICK               THE DAISY DUCKS
      6     BURKE, JAMES LEE          NEON RAIN
      7     BURNS, REX                THE KILLING ZONE
      8     CAMPBELL, ROBERT          HIP-DEEP IN ALLIGATORS
      9     CAUNITZ, WILLIAM J.       BLACK SAND
     10     COLLINS, MICHAEL          CASTRATO
     11     DOBYNS, STEPHEN           SARATOGA HEADHUNTER
     12     FREEMANTLE, BRIAN         THE CHOICE OF EDDIE FRANKS
     13     FRIEDMAN, BRUCE JAY       THE DICK
     14     GAULT, WILLIAM CAMPBELL   THE CHICANO WAR
     15     GILBERT, MICHAEL          PAINT, GOLD & BLOOD
     16     GILBERT, MICHAEL          TROUBLE
     17     GRAY, MALCOLM             A MATTER OF RECORD
     18     HEALY, JEREMIAH           THE STAKED GOAT
     19     HILLERMAN, TONY           COYOTE WAITS
     20     HILLERMAN, TONY           TALKING GOD
     21     KENNEALY, JERRY           POLO ANYONE
     22     KNOTT, FREDERICK          DIAL "M" FOR MURDER
     23     LATHEN, EMMA              SOMETHING IN THE AIR
     24     LEWIS, ROY HARLEY         MIRACLES TAKE A LITTLE LONGER
     25     LINDSEY, DAVID L.         HEAT FROM ANOTHER SUN
     26     LUTZ, JOHN                RIDE THE LIGHTNING
     27     LUTZ, JOHN                SHADOWTOWN
     28     LUTZ, JOHN                SHADOWTOWN
     29     LYONS, ARTHUR             OTHER PEOPLE'S MONEY
     30     MARSHALL, WILLIAM         HEAD FIRST
     31     MARSHALL, WILLIAM         THE FAR AWAY MAN
     32     MARSHALL, WILLIAM         WHISPER
     33     MAXWELL, A.E.             JUST ENOUGH LIGHT TO KILL
     34     MAXWELL, THOMAS           KISS ME ONCE
     35     MILNE, JOHN               THE MOODY MAN
     36     MULLER, MARCIA            THERE'S SOMETHING IN A SUNDAY
     37     Mac LEOD, CHARLOTTE       THE SILVER GHOST
```

```
OBS          AUTHOR                TITLE

 38    Mc BRIDE, BILL (ED.)    IDENTIFICATION OF FIRST EDITIONS
 39    Mc DONALD, JOHN D.      A KEY TO THE SUITE
 40    Mc DONALD, JOHN D.      NO DEADLY DRUG
 41    OLIVER, ANTHONY         COVER-UP
 42    PARKER, ROBERT B.       PALE KINGS AND PRINCES
 43    PARRISH, FRANK          DEATH IN THE RAIN
 44    RISENHOOVER, C.C.       CHILD STALKER
 45    SANGSTER, JIMMY         SNOWBALL
 46    SIMON, ROGER L.         RAISING THE DEAD
 47    SINGER, SHELLY          SPIT IN THE OCEAN
 48    SPENCER, ROSS H.        DEATH WORE GLOVES
 49    STARK, RICHARD          THE MOURNER
 50    TURNBULL, PETER         CONDITION PURPLE
 51    UHNAK, DOROTHY          VICTIMS
 52    UHNAK, DOROTHY          VICTIMS
 53    VACHSS, ANDREW          HARD CANDY
 54    WALKER, WALTER          THE IMMEDIATE PROSPECT OF BEING HANGED
 55    WALKER, WALTER          THE IMMEDIATE PROSPECT OF BEING HANGED
 56    WALKER, WALTER          THE TWO DUDE DEFENSE
 57    WESTLAKE, DONALD        SACRED MONSTER
 58    WILLEFORD, CHARLES      THE WAY WE DIE NOW
```

Solution: BY Statement, FIRST. and LAST. Variables

Each observation in SLUETH.BOOKS represents one book. The variable AUTHOR holds the author's name. When there is one and only one observation for a particular value of AUTHOR, it means there is just one book remaining in inventory for that author. You can use the temporary DATA step variables FIRST.*variable* and LAST.*variable* to find these unique, one-and-only-one values. FIRST.*variable* and LAST.*variable* variables are created for you automatically when you use a BY statement with a SET statement.

The name of the BY statement variable is appended to "FIRST." and "LAST." to form the full name of the temporary variables. In Program 12, the BY variable is AUTHOR so the temporary variables are FIRST.AUTHOR and LAST.AUTHOR. These variables are tested with a subsetting IF statement (see line 4) to determine if an author name is the first in the BY group **and** the last in the BY group, that is, it is the **only** one in the BY group.

FIRST. and LAST. temporary variables and the subsetting IF statement are distinctive features of the SAS DATA step language. See End Notes for more details.

Program 12

```
1   DATA LASTONE;
2      SET SLUETH.BOOKS;
3         BY AUTHOR;
4      IF FIRST.AUTHOR=1 AND LAST.AUTHOR=1;
5   RUN;
6
7   TITLE 'Only One Book Left by These Authors';
8
9   PROC PRINT DATA=LASTONE;
10  RUN;
```

Closer Look

There can be only one BY statement for each SET statement. The BY statement should be placed immediately after its associated SET statement. When used with the SET statement, the BY statement has one effect: the creation of the FIRST.*variable* and LAST.*variable* temporary variables.

Program 12 Notes

Line 1 The DATA statement starts the DATA step and names LASTONE as the output data set. LASTONE is a temporary SAS data set.

2 The SET statement reads each observation in SLEUTH.BOOKS.

3 When used with a SET statement in a SAS DATA step, the BY statement creates two temporary variables: FIRST.*variable* and LAST.*variable*. In this case, the variables are FIRST.AUTHOR and LAST.AUTHOR.

When you use a BY statement, the input SAS data set must be in BY-variable order. SLEUTH.BOOKS is assumed to be properly sorted.

4 This subsetting IF statement can be interpreted as follows: *If the current author name is the first in the BY group and the last in the BY group, then output to the SAS data set LASTONE; otherwise discard the current observation and continue processing with the next observation.* FIRST.AUTHOR is automatically set to 1 when the current value of AUTHOR is the first in a BY group; otherwise it is set to 0. LAST.AUTHOR is automatically set to 1 when the current value of AUTHOR is the last in a BY group; otherwise it is set to 0.

The effect of this statement is to select books by authors that appear once and only once in SLEUTH.BOOKS. This statement takes advantage of some unique DATA step features. See End Notes for details on the subsetting IF statement and FIRST. and LAST. temporary variables.

5 The RUN statement ends and executes the DATA step.

7 The TITLE statement applies to the PROC PRINT output. The quoted text will be displayed at the top of each page.

9 The PROC statement starts the PRINT procedure and names the temporary data set LASTONE as input.

10 The RUN statement ends and executes the PRINT procedure.

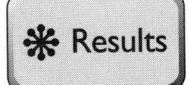

Results: List of Last Books by an Author

The low inventory report shown below is the output from the PROC PRINT step (lines 9-10). These were the only observations from the inventory data set SLEUTH.BOOKS that met the subsetting IF condition on line 4.

Authors with only one book in inventory

```
                  Only One Book Left By These Authors

     OBS    AUTHOR                    TITLE

     1      BARNES, LINDA             THE SNAKE TATTOO
     2      BURKE, JAMES LEE          NEON RAIN
     3      BURNS, REX                THE KILLING ZONE
     4      CAMPBELL, ROBERT          HIP-DEEP IN ALLIGATORS
     5      CAUNITZ, WILLIAM J.       BLACK SAND
     6      COLLINS, MICHAEL          CASTRATO
     7      DOBYNS, STEPHEN           SARATOGA HEADHUNTER
     8      FREEMANTLE, BRIAN         THE CHOICE OF EDDIE FRANKS
     9      FRIEDMAN, BRUCE JAY       THE DICK
     10     GAULT, WILLIAM CAMPBELL   THE CHICANO WAR
     11     GRAY, MALCOLM             A MATTER OF RECORD
     12     HEALY, JEREMIAH           THE STAKED GOAT
     13     KENNEALY, JERRY           POLO ANYONE
     14     KNOTT, FREDERICK          DIAL "M" FOR MURDER
     15     LATHEN, EMMA              SOMETHING IN THE AIR
     16     LEWIS, ROY HARLEY         MIRACLES TAKE A LITTLE LONGER
     17     LINDSEY, DAVID L.         HEAT FROM ANOTHER SUN
     18     LYONS, ARTHUR             OTHER PEOPLE'S MONEY
     19     MAXWELL, A.E.             JUST ENOUGH LIGHT TO KILL
     20     MAXWELL, THOMAS           KISS ME ONCE
     21     MILNE, JOHN               THE MOODY MAN
     22     MULLER, MARCIA            THERE'S SOMETHING IN A SUNDAY
     23     Mac LEOD, CHARLOTTE       THE SILVER GHOST
     24     Mc BRIDE, BILL (ED.)      IDENTIFICATION OF FIRST EDITIONS
     25     OLIVER, ANTHONY           COVER-UP
     26     PARKER, ROBERT B.         PALE KINGS AND PRINCES
     27     PARRISH, FRANK            DEATH IN THE RAIN
     28     RISENHOOVER, C.C.         CHILD STALKER
     29     SANGSTER, JIMMY           SNOWBALL
     30     SIMON, ROGER L.           RAISING THE DEAD
     31     SINGER, SHELLY            SPIT IN THE OCEAN
     32     SPENCER, ROSS H.          DEATH WORE GLOVES
     33     STARK, RICHARD            THE MOURNER
     34     TURNBULL, PETER           CONDITION PURPLE
     35     VACHSS, ANDREW            HARD CANDY
     36     WESTLAKE, DONALD          SACRED MONSTER
     37     WILLEFORD, CHARLES        THE WAY WE DIE NOW
```

Subsetting IF, FIRST. and LAST., Unique Values

FIRST. and LAST. variables. When you use a BY statement in a SAS DATA step, the temporary variables FIRST.*variable* and LAST.*variable* are automatically created. These are Boolean variables, meaning they have one of two possible values: 1 (true) or 0 (false). The values of FIRST.AUTHOR and LAST.AUTHOR for the first 30 observations processed in Program 12 appear below. You can see that both FIRST.AUTHOR and LAST.AUTHOR are set to 1 only when there is just one book by the author in inventory. FIRST.*variable* and LAST.*variable* values are set automatically by the SAS System.

FIRST.AUTHOR and LAST.AUTHOR for the first 30 observations

AUTHOR	TITLE	FIRST.AUTHOR	LAST.AUTHOR
ALLBEURY, TED	THE LANTERN NETWORK	1	0
ALLBEURY, TED	THE LANTERN NETWORK	0	1
BARNES, LINDA	THE SNAKE TATTOO	1	1
BOYER, RICK	THE DAISY DUCKS	1	0
BOYER, RICK	THE DAISY DUCKS	0	1
BURKE, JAMES LEE	NEON RAIN	1	1
BURNS, REX	THE KILLING ZONE	1	1
CAMPBELL, ROBERT	HIP-DEEP IN ALLIGATORS	1	1
CAUNITZ, WILLIAM J.	BLACK SAND	1	1
COLLINS, MICHAEL	CASTRATO	1	1
DOBYNS, STEPHEN	SARATOGA HEADHUNTER	1	1
FREEMANTLE, BRIAN	THE CHOICE OF EDDIE FRANKS	1	1
FRIEDMAN, BRUCE JAY	THE DICK	1	1
GAULT, WILLIAM CAMPBELL	THE CHICANO WAR	1	1
GILBERT, MICHAEL	PAINT, GOLD & BLOOD	1	0
GILBERT, MICHAEL	TROUBLE	0	1
GRAY, MALCOLM	A MATTER OF RECORD	1	1
HEALY, JEREMIAH	THE STAKED GOAT	1	1
HILLERMAN, TONY	COYOTE WAITS	1	0
HILLERMAN, TONY	TALKING GOD	0	1
KENNEALY, JERRY	POLO ANYONE	1	1
KNOTT, FREDERICK	DIAL "M" FOR MURDER	1	1
LATHEN, EMMA	SOMETHING IN THE AIR	1	1
LEWIS, ROY HARLEY	MIRACLES TAKE A LITTLE LONGER	1	1
LINDSEY, DAVID L.	HEAT FROM ANOTHER SUN	1	1
LUTZ, JOHN	RIDE THE LIGHTNING	1	0
LUTZ, JOHN	SHADOWTOWN	0	0
LUTZ, JOHN	SHADOWTOWN	0	1
LYONS, ARTHUR	OTHER PEOPLE'S MONEY	1	1
MARSHALL, WILLIAM	HEAD FIRST	1	0

Subsetting IF. The subsetting IF statement is a unique feature of the SAS DATA step language. Notice how it is different from the standard IF statement–there is no THEN! The subsetting IF has implied THEN and ELSE actions. The statement `IF X=1;` can be interpreted as follows:

If X=1 then proceed
else go immediately to the top of the DATA step.

To understand what this means, you have to keep in mind that the DATA step is a loop of executable statements with an implied write to the output SAS data set at the bottom of the loop. The program goes around and around the loop until there are no more observations to read. When the subsetting IF condition is true, the DATA step proceeds in its loop to the next statement. You can think of Program 12 as being made up of the following instructions:

1. SET SLUETH.BOOKS *(read an observation)*.
2. *If no more data then end the DATA step.*
3. *If FIRST.AUTHOR and LAST.AUTHOR both have a value of 1, then proceed.*
4. *Else, go to instruction 1 now.*
5. *Write an observation to data set LASTONE.*
6. *Go to the top of the DATA step, instruction 1.*

Since the subsetting IF is the last statement in Program 12, allowing the program to proceed means allowing the DATA step to write the current observation to the data set LASTONE then jump back to the top of the loop to read the next observation. When the condition is false, the program jumps immediately back to the top of the DATA step loop, bypassing any output to LASTONE. The only observations that get written to the output data set are those that satisfy the subsetting IF condition.

Filtering data. The subsetting IF statement acts as a filter. Each observation read by the SET statement is examined. Only observations that satisfy the IF condition continue in the DATA step loop. The WHERE statement also acts as a filter. For example, if you want to select all Fords from a used car inventory data set called CARS, you could use the subsetting IF as follows:

```
DATA FORDS;
   SET CARS;
   IF MAKE='Ford';
RUN;
```

Or you could use the WHERE statement:

```
DATA FORDS;
   SET CARS;
   WHERE MAKE='Ford';
RUN;
```

Unlike the subsetting IF, the WHERE statement filters observations **before** they are read by the SET statement. Observations that do not satisfy the WHERE condition never even make it into the DATA step loop. If all you need to do is filter data, the WHERE statement can be more efficient than the subsetting IF. Program 12 is a case where you must use the subsetting IF statement because the FIRST. and LAST. variables are not visible to the WHERE statement. For more information see "Understanding When to Choose WHERE Processing or the Subsetting IF Statement" in Chapter 14, "Reporting on Subsets of SAS Data Sets," *SAS Language and Procedures: Usage 2, Version 6, First Edition* and "WHERE" in Chapter 9, "SAS Language Statements," *SAS Language: Reference, Version 6, First Edition*.

Find unique values. In Program 12, you used the FIRST. and LAST. variables to select one-and-only-one authors from the SLEUTH.BOOKS data set. You can also use these variables to generate a list of all unique authors:

```
DATA AUTHORS;
   SET SLEUTH.BOOKS;
      BY AUTHOR;
   IF FIRST.AUTHOR;
RUN;
```

This DATA step uses the subsetting IF to check for FIRST.AUTHOR only, not FIRST.AUTHOR and LAST.AUTHOR. The effect of this change is that all author names are written to the AUTHORS data set once, no matter how many times they may appear in the data set. Check the values of FIRST.AUTHOR in the listing in the first End Note to convince yourself.

Sometimes you may want to select nonunique values from a data set. This is the logical opposite of the subsetting IF condition in Program 12. The program below selects all authors with two or more books in inventory:

```
DATA NOTLAST;
   SET SLEUTH.BOOKS;
      BY AUTHOR;
   IF NOT (FIRST.AUTHOR AND LAST.AUTHOR);
RUN;
```

Finding Unique Values

1. Make sure you know the names and have access to the SAS data sets you want to read. This may require a LIBNAME statement. (See Chapter 2 in this book for a LIBNAME example.)

2. Use a DATA step with SET and BY statements to read observations from the data set and create temporary FIRST.*variable* and LAST.*variable* variables.

3. Use the subsetting IF statement to test for FIRST.*variable*=1 **and** LAST.*variable*=1 to select one-and-only-one values.

DATA step, BY and IF Statements

BY statement, FIRST.*variable* and LAST.*variable* variables
"Listing Duplicate Values" in Chapter 15, "Producing Exception Reports," *SAS Language and Procedures: Usage 2, Version 6, First Edition;*
"BY-Group Processing" in Chapter 4, "Rules of the SAS Language," and "BY" in Chapter 9, "SAS Language Statements," *SAS Language: Reference, Version 6, First Edition*

DATA step
"How the DATA Step Works: a Basic Introduction" in Chapter 2, "Introduction to DATA Step Processing," *SAS Language and Procedures: Usage, Version 6, First Edition*;
Chapter 2, "The DATA Step," *SAS Language: Reference*

subsetting IF statement
"Getting a Total for Each BY Group" in Chapter 11, "Using More than One Observation in a Calculation," *SAS Language and Procedures: Usage;*
"IF" in Chapter 9, "SAS Language Statements," *SAS Language: Reference*

Chapter 13
Finding Percentages

Learn How to...

- Create a percentage report

Using These SAS System Features...

- FREQ procedure
- WEIGHT statement

? Problem — Calculate the Percent of Park Visits by Month

The Redcliff State Park tallies the number of visitors each month in SAS data set REDCLIFF.VISITS, shown below as PROC PRINT output. There is one observation for each month. The variable COUNT holds the total number of visitors for each month. To develop next year's budget, you need to create a report showing the percent of the annual total visits accounted for by each month.

SAS data set REDCLIFF.VISITS

OBS	MONTH	COUNT
1	JAN	152
2	FEB	140
3	MAR	278
4	APR	876
5	MAY	2578
6	JUN	5870
7	JUL	6563
8	AUG	6832
9	SEP	5027
10	OCT	2197
11	NOV	317
12	DEC	182

Solution | ## FREQ Procedure with the WEIGHT Statement

PROC FREQ counts the number of observations for each unique value of a variable. It also reports the percentage of all observations accounted for by those unique values (see Chapter 7 in this book). If you were to use PROC FREQ to count the number of unique values of the variable MONTH in data set REDCLIFF.VISITS, it would report one occurrence and a percentage of 8.3% (1 out of 12) for each month. The WEIGHT statement allows you to have PROC FREQ use the visitor count for each month instead of the number of times each month appears in the data set.

When you use the PROC FREQ statements **WEIGHT COUNT;** and **TABLE MONTH;** as in Program 13 on the following page, the number of occurrences of each unique value of MONTH (just 1) is multiplied by the value of the WEIGHT variable COUNT. That is, 1 times the count of visitors. The FREQ procedure uses this number to calculate occurrences and percentages. In effect, the WEIGHT statement says: *Pretend this observation occurs COUNT number of times instead of just once.* For the REDCLIFF.VISITS data set this means: *Pretend "JAN" occurs 152 times, "FEB" 140 times, "MAR" 278 times, etc.* The WEIGHT statement forces PROC FREQ to add up the number of visitors and report the percent of this total accounted for by each month.

Program 13

```
1   PROC FREQ DATA=REDCLIFF.VISITS ORDER=DATA;
2      TITLE 'Percentage of Redcliff Visitors By Month';
3      WEIGHT COUNT;
4      TABLE MONTH;
5   RUN;
```

Program 13 Notes

Line

1. The PROC statement begins the FREQ procedure and names REDCLIFF.VISITS as the input data set. The ORDER=DATA option tells PROC FREQ to report the percentages in the same order they occur in the input data set, that is, in month order from JAN to DEC.

2. The TITLE statement applies to the PROC FREQ output. The quoted text will be displayed at the top of each page of output.

3. The WEIGHT statement tells PROC FREQ to multiply the number of occurrences of the TABLE statement variable, MONTH, by the value of COUNT.

4. The TABLE statement tells PROC FREQ to report unique values of MONTH.

5. The RUN statement ends and executes the FREQ procedure.

Closer Look

If you leave off the **ORDER=DATA option** in Program 13, line 1, the output report would be in alphabetical month order: APR, AUG, DEC, etc. See "PROC FREQ Statement" in Chapter 20, "The FREQ Procedure," *SAS Procedures Guide, Version 6, Third Edition* for more on the ORDER= option.

132 Part 2 - Working with Your Data

✻ Results

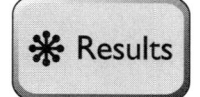

Monthly Percentage Report

The output from Program 13 appears below. Each unique value of the TABLE variable, MONTH, is reported. The `Frequency` column lists the count of each unique month (always 1) multiplied by the value of COUNT, as specified in the WEIGHT statement. `Frequency` is actually the number of visitors (1✕COUNT). `Percent` is the number you need to find. It represents the percent of all visits accounted for by each month. The cumulative columns show values for the current month plus all previous months. For example, you can see that the first four months of the year, JAN through APR, accounted for only 4.7% of all visits.

PROC FREQ output: percentages

```
              Percentage of Redcliff Visitors By Month

                                      Cumulative   Cumulative
        MONTH    Frequency    Percent  Frequency    Percent
        ------------------------------------------------------
        JAN          152        0.5        152        0.5
        FEB          140        0.5        292        0.9
        MAR          278        0.9        570        1.8
        APR          876        2.8       1446        4.7
        MAY         2578        8.3       4024       13.0
        JUN         5870       18.9       9894       31.9
        JUL         6563       21.2      16457       53.1
        AUG         6832       22.0      23289       75.1
        SEP         5027       16.2      28316       91.3
        OCT         2197        7.1      30513       98.4
        NOV          317        1.0      30830       99.4
        DEC          182        0.6      31012      100.0
```

🔍 Closer Look

With PROC FREQ you have **no direct control over the number of decimal places** reported in the `Percent` column. If you need more control over decimal places or other presentation qualities, you can send the PROC FREQ output to a SAS data set (see Chapter 5 in this book), then use PROC PRINT to generate your report.

Chapter 13 - Finding Percentages **133**

End Notes: More on the WEIGHT Statement and Percentages

More on the WEIGHT statement. The WEIGHT statement in PROC FREQ can be useful any time you need to sum values rather than count occurrences. For example, you need to put together an inventory of microcomputer software at your company. Microcomputer users send you a list of the software installed on their PCs, including the type and cost. The data are entered in SAS data set BIGCO.PCSOFT with the variables TYPE and COST:

SAS data set BIGCO.PCSOFT

```
OBS      TYPE          COST

 1    spreadsheet     295.98
 2    wordproc        279.65
 3    graphics        425.35
 4    spreadsheet     225.99
 5    spreadsheet     255.00
 6    database        485.03
 7    wordproc        275.00
 8    spreadsheet     298.69
```

In the program below, PROC FREQ with a WEIGHT statement generates a report showing how much was spent on each type of software and the percent of the total accounted for by each type. The output follows the program.

```
PROC FREQ DATA=BIGCO.PCSOFT;
    WEIGHT COST;
    TABLE TYPE;
RUN;
```

Software cost report

```
                                     Cumulative  Cumulative
TYPE          Frequency    Percent   Frequency   Percent
-----------------------------------------------------------
database         485.03      19.1       485.03      19.1
graphics         425.35      16.7       910.38      35.8
spreadsheet     1075.66      42.3      1986.04      78.2
wordproc         554.65      21.8      2540.69     100.0
```

In the output above, `Frequency` is **not** a count; it is the total cost for each type of software package. `Percent` tells you the percentage of all software expenditures accounted for by each product category. For example, you can see that spreadsheet products accounted for 42.3% of software expenses.

134 Part 2 - Working with Your Data

Note that if all you need are the sums, you're better off using the MEANS procedure because of its greater flexibility (see Chapter 8). The FREQ procedure with the WEIGHT statement is useful when you need both sums and percentages.

Percentages. There are other SAS System tools you can use to tackle the percentage problem. You could use PROC TABULATE, the DATA step, or a combination of DATA and PROC steps. (See Chapter 14 for more on PROC TABULATE.)

The DATA step language allows you to write procedural computer programs, that is, programs made up of sets of explicit instructions. This means you can do just about anything with a DATA step program, and that certainly includes calculating percentages. The general solution involves two operations:

1. Find the sum of all values.
2. Divide individual values by the sum, and multiply by 100.

In a DATA step you can read a SAS data set twice: once to get a total, then again to divide individual values by that total for the percentage.

Another strategy is to use PROC MEANS to generate the grand total and individual sums and write them to a SAS data set. The data set can then be read by a subsequent DATA step that performs the percentage division.

These methods are more complicated than PROC FREQ with the WEIGHT statement, but they do give you more control over formatting and data selection.

TABLE statement options. You can control the output from PROC FREQ with certain TABLE statement options. If you don't want cumulative statistics, use the NOCUM option:

```
PROC FREQ DATA=REDCLIFF.VISITS ORDER=DATA;
   TITLE 'Percentage of Redcliff Visitors By Month';
   WEIGHT COUNT;
   TABLE MONTH / NOCUM;
RUN;
```

TABLE statement options must be separated from TABLE statement variables by a "/".

For the park visit report in this chapter you really only need the `Percent` column of the PROC FREQ output, but PROC FREQ does not allow you to suppress the `Frequency` column.

Finding Percentages

1. Make sure you know the names and have access to the SAS data sets with which you want to work. This may require a LIBNAME statement. (See Chapter 2 in this book for a LIBNAME example.)

2. Use the FREQ procedure with a WEIGHT statement. You get a percentage calculated for each value of the variable named in the TABLE statement. Name the variable for which you want a percentage in the WEIGHT statement.

FREQ Procedure

FREQ procedure and WEIGHT statement
Chapter 20 "The FREQ Procedure," *SAS Procedures Guide, Version 6, Third Edition*

Part 3 - Presenting Your Data

Chapter 14 - Creating a Table

Chapter 15 - Creating a Custom Report

Chapter 16 - Creating Bar Charts

Chapter 17 - Creating Stacked Bar Charts

Chapter 18 - Creating Grouped Bar Charts

Chapter 19 - Creating Line Graphs and Plots

Chapter 14
Creating a Table

Learn How to...

- Create a table showing averages
- Control table layout and cell formats
- Group data for inclusion in a table

Using These SAS System Features...

- TABULATE procedure
- CLASS statement
- TABLE statement

? Problem — Create an Experimental Results Table

An experiment was conducted to test the interaction effects of two fertilizer solutions. The experiment tested six different concentrations of the two solutions. Each concentration of solution 1 is combined with each concentration of solution 2 yielding 36 (6×6) different combinations. The solutions were applied to plants in three separate trials. The growth rate of the plants was measured in centimeters per day and recorded in the SAS data set AGLAB.GROW1, which appears on the following page. The variable TRIAL indicates which of the three experimental trials the results come from. SOL1 indicates the concentration of fertilizer solution 1, and SOL2 is the concentration of solution 2. GROWRATE is the observed growth rate. There are 108 observations: one for each combination of fertilizer concentrations times 3 trials (6×6×3=108).

You need to create a two-dimensional table that presents the experimental results. Rows in this table should indicate the concentration of solution 1, and columns should indicate the concentration of solution 2. For each combination of solution 1 and solution 2, cells in the table should show the growth rate average of the three trials.

SAS data set AGLAB.GROW1

```
              Fertilizer Solution Experiment Results

       OBS      TRIAL      SOL1      SOL2     GROWRATE

         1         1        0.0       0.0       0.13
         2         2        0.0       0.0       0.40
         3         3        0.0       0.0       0.22
         4         1        0.0       0.1       0.30
         5         2        0.0       0.1       0.24
         6         3        0.0       0.1       0.88
         7         1        0.0       0.2       0.57
         8         2        0.0       0.2       0.06
         9         3        0.0       0.2       0.26
        10         1        0.0       0.3       0.43
        11         2        0.0       0.3       0.85
        12         3        0.0       0.3       0.72
        13         1        0.0       0.4       0.87
        14         2        0.0       0.4       0.83
        15         3        0.0       0.4       0.61
        16         1        0.0       0.5       0.32
        17         2        0.0       0.5       0.04
        18         3        0.0       0.5       0.59
        19         1        0.1       0.0       0.11
        20         2        0.1       0.0       0.64
        21         3        0.1       0.0       0.85
        22         1        0.1       0.1       0.71
        23         2        0.1       0.1       0.36
        24         3        0.1       0.1       0.34
                                  .
                                  .
                                  .
        87         3        0.4       0.4       0.06
        88         1        0.4       0.5       0.21
        89         2        0.4       0.5       0.81
        90         3        0.4       0.5       0.68
        91         1        0.5       0.0       0.68
        92         2        0.5       0.0       0.04
        93         3        0.5       0.0       0.91
        94         1        0.5       0.1       0.43
        95         2        0.5       0.1       0.59
        96         3        0.5       0.1       0.85
        97         1        0.5       0.2       0.60
        98         2        0.5       0.2       0.59
        99         3        0.5       0.2       0.30
       100         1        0.5       0.3       0.57
       101         2        0.5       0.3       0.66
       102         3        0.5       0.3       0.69
       103         1        0.5       0.4       0.74
       104         2        0.5       0.4       0.83
       105         3        0.5       0.4       0.94
       106         1        0.5       0.5       0.48
       107         2        0.5       0.5       0.48
       108         3        0.5       0.5       0.70
```

Solution: PROC TABULATE

The required table can be created with the TABULATE procedure. PROC TABULATE can generate one-, two-, or three-dimensional tables, giving you extensive control over the appearance and content of your table. Program 14 and Program Notes show how to use PROC TABULATE to generate the experimental results table.

Program 14

```
1  PROC TABULATE DATA=AGLAB.GROW1 FORMAT=4.2;
2     TITLE 'Growth Rate Table';
3     LABEL SOL1='Solution 1 Concentration'
4           SOL2='Solution 2 Concentration';
5     CLASS SOL1 SOL2;
6     VAR GROWRATE;
7     TABLE SOL1, SOL2*GROWRATE=' '*MEAN=' ';
8  RUN;
```

Program 14 Notes

Line 1 The PROC statement begins the TABULATE procedure and names AGLAB.GROW1 as the input data set. The FORMAT=4.2 option sets the default format for each cell in the table. The 4.2 format displays a maximum of four characters with two decimal places. 9.99 is the largest number that can be displayed.

2 The TITLE statement applies to the TABULATE procedure output. The quoted text will be displayed at the top of each page.

3 The LABEL statement assigns text to be used in place of the name of a variable. "Solution 1 Concentration" will appear in the table instead of the variable name "SOL1". "SOL2" is also replaced. You can assign one or more variable name substitutions in a single LABEL statement.

Closer Look

The LABEL statement is not specific to PROC TABULATE. It can be used with many procedures that generate printed output such as PROC PRINT, PROC FREQ, and PROC MEANS. The LABEL statement substitutes a string of text for a variable name. It is often used with PROC PRINT to create customized column headings. See "LABEL" in Chapter 9, "SAS Language Statements," *SAS Language: Reference, Version 6, First Edition* for details.

Program 14 Notes

Line

5 The CLASS statement tells PROC TABULATE to group data by the variables SOL1 and SOL2 regardless of the value of TRIAL. The MEAN statistic requested in line 9 will be calculated across all three trials for each combination of SOL1 and SOL2.

6 The VAR statement names the variable you want to analyze, in this case the variable for which you want to find the mean: GROWRATE.

7 The TABLE statement defines the layout and cell contents of the table. Tables can have one, two, or three dimensions. Dimensions are separated by commas. Here, two dimensions are defined. The first is SOL1. The six unique values of SOL1 make up the six rows of the table. The second dimension is defined by naming SOL2 after the comma. The six unique values of SOL2 make up the columns of the table.

The asterisk is called the crossing operator. It can be interpreted as follows: *Show this...* The asterisk after SOL2 tells the SAS System to *show a value for GROWRATE under each SOL2 column*. The "*MEAN" following the GROWRATE specification names the type of value you want to show. Here, it is the mean (average) of GROWRATE and can be interpreted as follows: *Show the mean of GROWRATE in each cell.*

The =' ' (equal sign followed by a single quote, blank, and single quote) after GROWRATE and MEAN are used to suppress the listing of subheadings. If you leave this out, the text "GROWRATE" and "MEAN" would appear as column subheadings. The equal sign modifier tells the SAS System to *use the following quoted text instead of "GROWRATE" or "MEAN"*. Here, the following text is a blank (' ') so subheadings are eliminated.

Every variable named in the TABLE statement must be listed in a preceding CLASS or VAR statement. A general rule is that if the values of the variable will be row, column, or page headings then you must name the variable in the CLASS statement. You must name values that you want to analyze (sum, average, etc.) within the CLASS headings in the VAR statement.

8 The RUN statement ends and executes the TABULATE procedure.

Closer Look

You can generate many other statistics with PROC TABULATE. In addition to mean, there is sum, N (count of nonmissing values), minimum, maximum, and several others. For a list of available statistics, see "Keywords and Formulas" in Chapter 37, "The TABULATE Procedure," *SAS Procedures Guide, Version 6, Third Edition*.

Average Growth Rate Table

The output from PROC TABULATE in Program 14 appears below. The TABLE statement specifies SOL1 as the first table dimension and SOL2 as the second dimension. The values of SOL1 form the row headings. SOL2 values are column headings. Each cell shows the mean of GROWRATE as specified by the crossing operators (*). Note the LABEL statement text is used instead of the variable names "SOL1" and "SOL2". Also note the format of the cell values: four total characters with two decimal places as specified by the PROC statement option FORMAT=4.2.

Growth rate table: output from PROC TABULATE

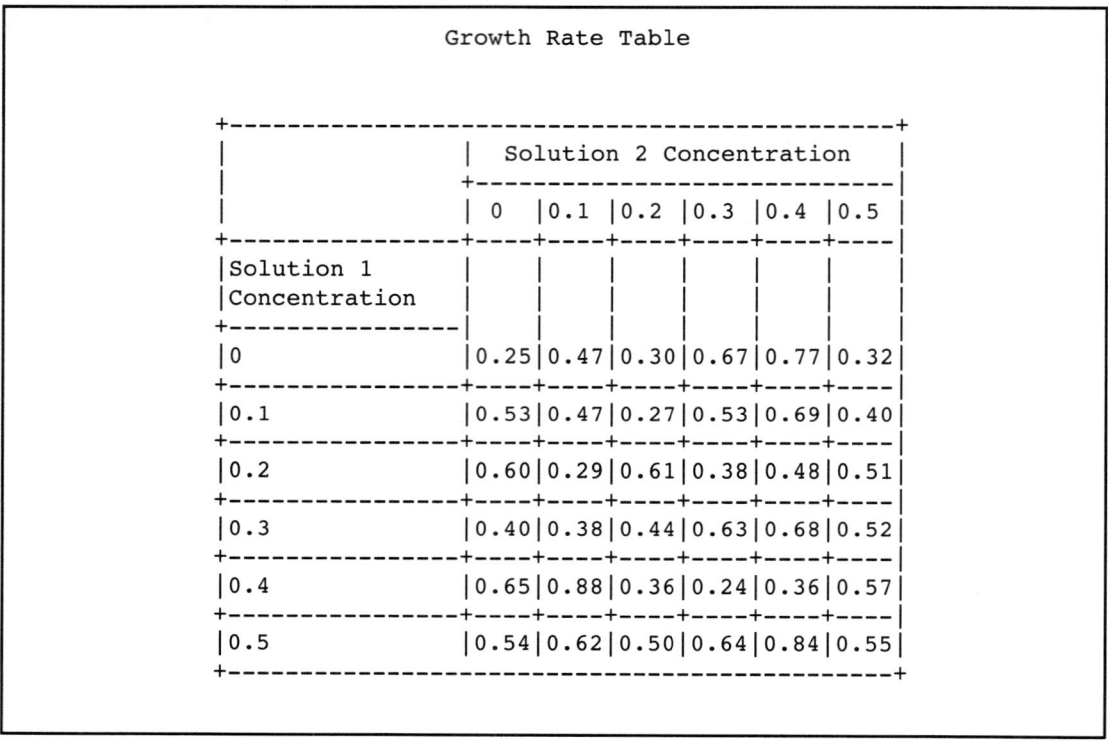

```
                                Growth Rate Table


          +-------------------------------------------------+
          |               |  Solution 2 Concentration       |
          |               +---------------------------------|
          |               | 0   |0.1 |0.2 |0.3 |0.4 |0.5   |
          +---------------+-----+----+----+----+----+----- |
          |Solution 1     |     |    |    |    |    |      |
          |Concentration  |     |    |    |    |    |      |
          +---------------|     |    |    |    |    |      |
          |0              |0.25 |0.47|0.30|0.67|0.77|0.32  |
          +---------------+-----+----+----+----+----+----- |
          |0.1            |0.53 |0.47|0.27|0.53|0.69|0.40  |
          +---------------+-----+----+----+----+----+----- |
          |0.2            |0.60 |0.29|0.61|0.38|0.48|0.51  |
          +---------------+-----+----+----+----+----+----- |
          |0.3            |0.40 |0.38|0.44|0.63|0.68|0.52  |
          +---------------+-----+----+----+----+----+----- |
          |0.4            |0.65 |0.88|0.36|0.24|0.36|0.57  |
          +---------------+-----+----+----+----+----+----- |
          |0.5            |0.54 |0.62|0.50|0.64|0.84|0.55  |
          +-------------------------------------------------+
```

TPL, the Crossing Operator, Table Dimensions

Table Producing Language. You should consider using PROC TABULATE if you routinely generate tables. The procedure allows you to group data, and calculate percentages and a variety of statistics, all in a single step and without DATA step programming. PROC TABULATE is one of the most powerful of all SAS procedures in terms of what you can do with just a few statements.

The TABLE statement syntax in PROC TABULATE is based on TPL (Table Producing Language) developed by the Bureau of Labor Statistics. TPL is made up of a series of operators: the asterisk (*) and the comma (,) as used in Program 14; plus the blank space concatenation operator and the parentheses associative operator, not used in Program 14. In addition there are = (equal sign) modifiers and other keywords that generate percentages and create summations. PROC TABULATE has additional features that give you extensive control over the appearance and content of your table. The following is a single TABLE statement taken from page 145 of the *SAS Guide to TABULATE Processing, Second Edition*:

```
TABLE SITE='Clinic location' ALL='All clinics',
  (PROV1=' ' ALL='Summary')*
  (N='# of visits'*F=6. TOTBILL='Worth of Services'*(MEAN SUM
  PCTSUM<SITE*PROV1*TOTBILL
      SITE*ALL*TOTBILL
      ALL*PROV1*TOTBILL
      ALL*TOTBILL>='% of worth to worth of all services')
  TOTDUE='Amount Billed'*
  (SUM PCTSUM<TOTBILL>='% of worth'*F=7.2)) / RTS=12 CON-
DENSE;
```

This is not meant to scare you away from PROC TABULATE, but be aware that because TPL is so compact it can be baffling when you first use it. Be prepared to experiment a little.

> **The crossing operator.** In the TABLE statement, the asterisk (*) is called the crossing operator. You can interpret it as *show this...*

SOL2*GROWRATE means: *Show GROWRATE for each value of SOL2.*
SOL2*GROWRATE*MEAN means: *Show the mean of GROWRATE for each SOL2.*

If you remove all crossing operators from Program 14 and use the following TABLE statement, you can get the table that appears in the output below:

TABLE SOL1, SOL2;

Table output after removing the crossing operator

```
                          Growth Rate Table
                     Fertilizer Solution Experiment
   +----------------------------------------------------------+
   |                      |       Solution 2 Concentration    |
   |                      +-----------------------------------|
   |                      | 0   |0.1 |0.2 |0.3 |0.4 |0.5 |
   |                      +----+----+----+----+----+----|
   |                      | N  | N  | N  | N  | N  | N  |
   +----------------------+----+----+----+----+----+----|
   |Solution 1            |    |    |    |    |    |    |
   |Concentration         |    |    |    |    |    |    |
   +----------------------|    |    |    |    |    |    |
   |0                     |3.00|3.00|3.00|3.00|3.00|3.00|
   +----------------------+----+----+----+----+----+----|
   |0.1                   |3.00|3.00|3.00|3.00|3.00|3.00|
   +----------------------+----+----+----+----+----+----|
   |0.2                   |3.00|3.00|3.00|3.00|3.00|3.00|
   +----------------------+----+----+----+----+----+----|
   |0.3                   |3.00|3.00|3.00|3.00|3.00|3.00|
   +----------------------+----+----+----+----+----+----|
   |0.4                   |3.00|3.00|3.00|3.00|3.00|3.00|
   +----------------------+----+----+----+----+----+----|
   |0.5                   |3.00|3.00|3.00|3.00|3.00|3.00|
   +----------------------------------------------------------+
```

When you don't specify a *show this...* crossing operator, PROC TABULATE shows the N statistic. N is the count of observations with nonmissing values. The default *show this...* is *show the count*. The table above tells you there were three observations where SOL1 was 0.2 and SOL2 was 0.1. In fact, there were three observations for all combinations of SOL1 and SOL2, one from each of the three trials.

📝 **Table dimensions.** PROC TABULATE can create one-, two-, or three-dimensional tables (columns; rows and columns; or rows, columns, and pages). Dimensions are determined by the comma operator in the TABLE statement. A TABLE statement with no commas generates a one-dimensional table with columns only. In the following example there are no commas in the TABLE statement:

```
PROC TABULATE DATA=AGLAB.GROW1 FORMAT=4.2;
   TITLE 'One Dimension';
   CLASS SOL1;
   VAR GROWRATE;
   TABLE SOL1*GROWRATE=' '*MEAN=' ';
RUN;
```

The resulting table has one dimension: columns.

A one-dimensional, columns-only table

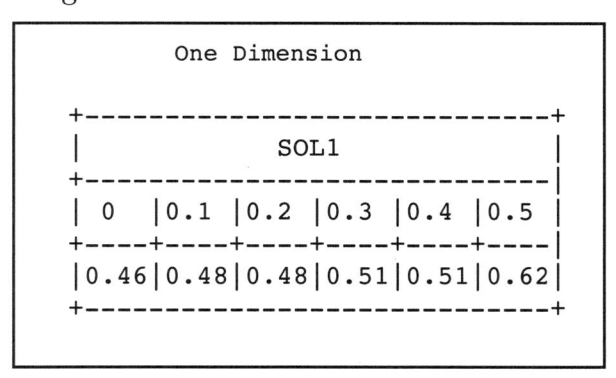

```
             One Dimension

    +-------------------------------+
    |             SOL1              |
    +-------------------------------|
    |  0 |0.1 |0.2 |0.3 |0.4 |0.5  |
    +----+----+----+----+----+----|
    |0.46|0.48|0.48|0.51|0.51|0.62|
    +-------------------------------+
```

There are six columns, one for each unique value of SOL1. The cells represent the mean of all 18 values (3 trials times 6 SOL2 values) of GROWRATE for each value of SOL1.

Chapter 14 - Creating a Table **147**

Creating a Table

1. Make sure you know the name and have access to the SAS data set with which you want to work. This may require a LIBNAME statement. (See Chapter 2 in this book for a LIBNAME example.)

2. Use PROC TABULATE. Name the variables that will act as column and row headings in the CLASS statement.

3. Name the variables that you want to analyze (sum, average, etc.) in the VAR statement.

4. Use the TABLE statement to define the structure of your table. Commas separate dimensions. The * means *show this...* Name statistics keywords after the variables you want them to apply to with an * between the variable name and the statistic name.

The TABULATE Procedure

TABULATE procedure, general reference
Chapter 37, "The TABULATE Procedure," *SAS Procedures Guide, Version 6, Third Edition*

TABULATE procedure, in-depth reference and tutorials
SAS Guide to TABULATE Processing, Second Edition

TABULATE procedure, usage and examples
Chapter 25, "Creating Summary Tables," *SAS Language and Procedures: Usage, Version 6, First Edition*

Also see keyword listings under PROC TABULATE in the annual SUGI proceedings. SUGI is the international SAS software users group. SUGI publishes papers presented during each conference. Over the years, several papers and tutorials have been presented dealing with PROC TABULATE. SUGI proceedings are available from SAS Institute's Book Sales Dept.

Chapter 15
Creating a Custom Report

Learn How to...

- Develop a custom report layout
- Handle data groups (control breaks)
- Accumulate totals and counts
- Control page headings

Using These SAS System Features...

- DATA step
- SET and BY statements
- FILE statement and options
- PUT statement
- LINK and RETURN statements

? Problem Generate an Auto Insurance Claims Report

The claims processing department keeps a listing of all auto accident insurance claims in SAS data set AUTODIV.CLAIMS, which appears on the next page. The date of the claim (DATE), injury category (CATEGORY), sales region (REGION), and amount of the claim (AMOUNT) are recorded. The variable DATE is shown in the DATE7. format.

You need to create a report listing all claims by category in descending order of claim amount. The dollar total and number of claims in each category must appear. The layout for the final claims report is as follows:

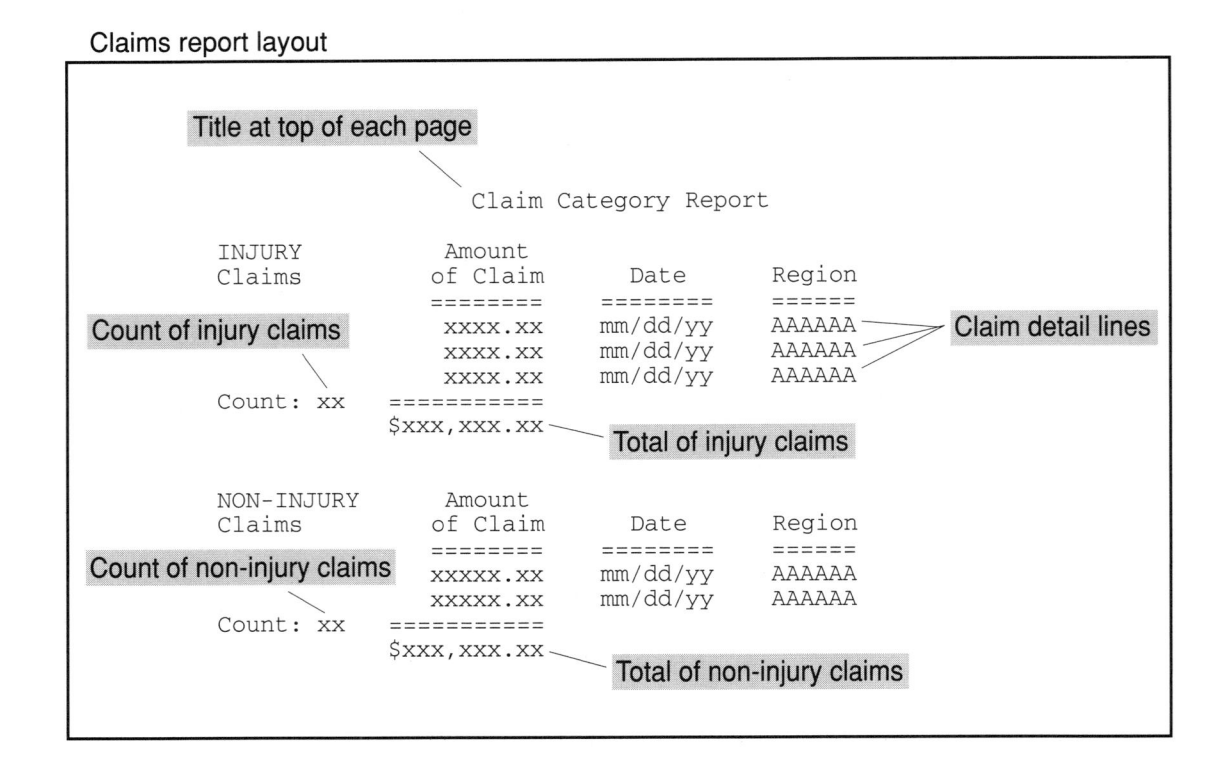

Claims report layout

```
                         Claim Category Report        ← Title at top of each page

          INJURY         Amount
          Claims         of Claim      Date        Region
                         ========      ========    ======
                         xxxx.xx       mm/dd/yy    AAAAAA  ⎫
                         xxxx.xx       mm/dd/yy    AAAAAA  ⎬ Claim detail lines
                         xxxx.xx       mm/dd/yy    AAAAAA  ⎭
          Count: xx      ===========
                         $xxx,xxx.xx   ← Total of injury claims

          NON-INJURY     Amount
          Claims         of Claim      Date        Region
                         ========      ========    ======
                         xxxxx.xx      mm/dd/yy    AAAAAA
                         xxxxx.xx      mm/dd/yy    AAAAAA
          Count: xx      ===========
                         $xxx,xxx.xx   ← Total of non-injury claims
```

Count of injury claims (points to Count: xx under INJURY)

Count of non-injury claims (points to Count: xx under NON-INJURY)

SAS data set AUTODIV.CLAIMS

```
                           Claims List
       OBS     DATE      CATEGORY       REGION      AMOUNT

        1    04JAN93    INJURY         NORTH      14454.19
        2    09JAN93    NON-INJURY     SOUTH       6362.79
        3    09JAN93    INJURY         NORTH      10681.64
        4    10JAN93    INJURY         SOUTH      10780.58
        5    12JAN93    NON-INJURY     NORTH       8375.73
        6    19JAN93    NON-INJURY     WEST        7838.12
        7    23FEB93    INJURY         SOUTH       7797.22
        8    26FEB93    INJURY         SOUTH       3743.85
        9    27FEB93    NON-INJURY     NORTH       9515.27
       10    01MAR93    NON-INJURY     WEST        7830.25
       11    04MAR93    NON-INJURY     WEST       10066.23
       12    15MAR93    NON-INJURY     EAST        2917.64
       13    19MAR93    INJURY         EAST        6447.74
       14    07APR93    INJURY         SOUTH      16679.13
       15    15APR93    INJURY         WEST       18963.10
       16    17APR93    INJURY         EAST        5248.55
       17    21APR93    NON-INJURY     EAST        3785.98
       18    24APR93    INJURY         SOUTH       9060.88
       19    28APR93    INJURY         EAST       16837.60
       20    28APR93    INJURY         NORTH      19551.32
       21    03MAY93    NON-INJURY     SOUTH       7846.33
       22    11MAY93    NON-INJURY     SOUTH      11434.47
       23    14MAY93    NON-INJURY     SOUTH       9140.13
       24    18MAY93    NON-INJURY     EAST       11231.73
       25    20MAY93    NON-INJURY     SOUTH      11462.81
       26    21MAY93    NON-INJURY     SOUTH      19748.66
       27    27MAY93    NON-INJURY     NORTH      11269.33
       28    08JUN93    INJURY         SOUTH      18264.37
       29    10JUN93    NON-INJURY     WEST        1026.76
       30    12JUN93    NON-INJURY     WEST        3015.00
       31    14JUN93    NON-INJURY     EAST        5814.65
       32    16JUN93    NON-INJURY     WEST        1468.71
       33    29JUN93    NON-INJURY     SOUTH      12661.89
       34    01JUL93    NON-INJURY     SOUTH       3179.99
       35    02JUL93    NON-INJURY     EAST        5544.70
       36    03JUL93    NON-INJURY     NORTH       8191.58
       37    11JUL93    NON-INJURY     NORTH       1573.88
       38    14JUL93    NON-INJURY     EAST       18539.09
       39    16JUL93    NON-INJURY     WEST        7135.95
       40    20JUL93    NON-INJURY     SOUTH       9984.92
       41    28JUL93    NON-INJURY     WEST       15822.52
       42    03AUG93    INJURY         EAST       16533.09
       43    11AUG93    NON-INJURY     NORTH       4628.15
       44    18AUG93    NON-INJURY     EAST       10137.39
       45    28AUG93    NON-INJURY     WEST        7422.62
       46    28AUG93    NON-INJURY     SOUTH      16091.38
       47    08SEP93    NON-INJURY     SOUTH       6281.82
       48    08SEP93    NON-INJURY     NORTH      14135.71
       49    17SEP93    NON-INJURY     NORTH       9515.56
       50    27SEP93    NON-INJURY     EAST       19511.68
       51    29SEP93    NON-INJURY     WEST        6643.65
       52    05OCT93    NON-INJURY     NORTH      19770.49
       53    11OCT93    NON-INJURY     WEST        9971.64
       54    19OCT93    NON-INJURY     WEST        3284.94
       55    05NOV93    NON-INJURY     WEST        5913.21
       56    14NOV93    INJURY         WEST       13966.44
       57    29NOV93    INJURY         NORTH      19225.54
       58    30NOV93    NON-INJURY     WEST        2024.45
       59    04DEC93    NON-INJURY     NORTH      19829.24
       60    05DEC93    NON-INJURY     SOUTH      15316.09
       61    08DEC93    NON-INJURY     SOUTH       7823.31
       62    14DEC93    NON-INJURY     EAST        5279.15
       63    19DEC93    NON-INJURY     SOUTH       5350.47
       64    23DEC93    NON-INJURY     EAST        1129.14
       65    24DEC93    NON-INJURY     SOUTH      17991.21
```

Solution: DATA Step Report Writing

You can generate the required report with the SAS DATA step. DATA step statements let you read observations, accumulate totals and counts, and control the start of a new page (page break). The solution in Program 15 consists of two SAS steps. First, the AUTODIV.CLAIMS data set is grouped by injury category (CATEGORY) and arranged by descending claim amount using a SORT procedure step. Then, the report is written with the DATA step.

One approach to DATA step report writing is to make a list of the things you need to do. For example, you may need to accumulate totals, print a header at the top of each new page, determine when a data group changes, etc. You can match this list to DATA step statements and options. See End Notes for more on planning your DATA step report and a table of reporting requirements and DATA step features that can help you get started.

Program 15

```
1   PROC SORT DATA=MYLIB.CLAIMS;
2      BY CATEGORY DESCENDING AMOUNT;
3   RUN;
4
5   DATA _NULL_;
6
7      TITLE 'Claim Category Report';
8      FILE PRINT LL=LINELEFT;
9
10     SET MYLIB.CLAIMS;
11         BY CATEGORY;
12
13     IF FIRST.CATEGORY = 1 THEN DO;
14         IF LINELEFT < 8 THEN PUT _PAGE_;
15         LINK CAT_HDR;
16         CAT_SUM = 0;
17         CAT_CNT = 0;
18     END;
19
20     CAT_SUM + AMOUNT;
21     CAT_CNT + 1;
22
```

152 Part 3 - Presenting Your Data

Program 15 (continued)

```
23        IF LINELEFT < 3 THEN DO;
24           PUT _PAGE_;
25           LINK CAT_HDR;
26        END;
27
28        PUT @15 AMOUNT 9.2
29            @28 DATE MMDDYY8.
30            @41 REGION;
31
32        IF LAST.CATEGORY = 1 THEN LINK SUM_CAT;
33
34        RETURN;
35
36     CAT_HDR:
37        PUT / /
38            @1  CATEGORY
39            @17 'Amount'
40            /
41            @1  'Claims'
42            @16 'of Claim'
43            @30 'Date'
44            @40 'Region'
45            /
46            @16 '========'
47            @28 '========'
48            @40 '======';
49        RETURN;
50
51     SUM_CAT:
52        PUT @1   'Count: ' CAT_CNT
53            @13  '==========='
54            /
56            @11  CAT_SUM DOLLAR13.2;
57        RETURN;
58
59  RUN;
```

Program 15 Notes

Line 1   The PROC statement begins the SORT procedure and names AUTODIV.CLAIMS as the input data set. No output data set is named so the original AUTODIV.CLAIMS will be replaced by the new, sorted version. (See Chapter 6 in this book for more on PROC SORT.)

2   The BY statement names the variable to sort on. AUTODIV.CLAIMS will be sorted first by claim category, then within each category by the descending amount of the claim. You can place the DESCENDING keyword in front of any variable in a BY statement to reverse the sort order.

3   The RUN statement ends and executes the SORT procedure.

5   The DATA statement begins the DATA step. "_NULL_" is a special name that means no SAS data set will be created in this DATA step. You need to create a printed report, not a SAS data set.

7   The TITLE statement applies to the DATA step output. The quoted text will appear at the top of each page.

8   The FILE statement names the output file for subsequent PUT statements. Here, the special file name PRINT is used. PRINT is the standard SAS print output target for your system. It may be a real printer or a file you can view online. If you are using the SAS Display Manager System, PRINT is the Output window. The PRINT destination is the same place that PROC PRINT output would go.

The LL= option says: *Keep track of how many lines are left on the page and keep that number in a SAS variable.* Here, that variable is LINELEFT. The name "LINELEFT" is arbitrary; you could use any valid SAS variable name.

10   The SET statement reads one observation at a time from SAS data set AUTODIV.CLAIMS.

11   This BY statement is associated with the preceding SET statement. When you use a BY statement with a SET statement, FIRST.*variable* and LAST.*variable* temporary variables are created. In this case, these variables are FIRST.CATEGORY and LAST.CATEGORY. (See Chapter 12 in this book for more on these temporary variables.)

FIRST.CATEGORY will be equal to 1 when the value of CATEGORY in the current observation is different than in the last observation. In other words, it is the **first** of a new injury category group. Otherwise, FIRST.CATEGORY will be 0. LAST.CATEGORY will be equal to 1 when the value of CATEGORY in the next observation is different than the current value. Otherwise, it is 0. When the value of LAST.CATEGORY is 1, the current observation is the **last** of an injury category group.

Program 15 Notes

Line

13 The IF statement tests the value of FIRST.CATEGORY. It asks: *Is the current value of CATEGORY the first in the BY group?* If it is, the statements in the DO group (lines 14-17) are executed, otherwise control skips to line 20. The statements in the DO group handle the printing of header lines.

14 This IF statement checks the value of LINELEFT, the variable named in the FILE statement LL= option on line 8. If there are less than eight lines left on the page, then the PUT _PAGE_ statement is executed, forcing output to a new page. Why test for eight lines? You always need enough room left on the page to print the title (one line), the header (four lines), a detail line (one line), and the summary (two lines). If you don't allow this much room, a header or summary could split across pages.

15 The LINK statement causes control to jump to the CAT_HDR label at line 36. The statements following CAT_HDR print the header lines.

16 The sum of claim amounts for an injury category is accumulated in the variable CAT_SUM. Here, CAT_SUM is reset to 0 in this DO group because you are dealing with a new injury category (FIRST. CATEGORY = 1).

17 The variable CAT_CNT accumulates the count of observations for each injury category. Like CAT_SUM, it is reset to 0 in this DO group because this is a new injury category (FIRST. CATEGORY = 1).

18 You must terminate all DO groups with an END statement. This END statement terminates the DO group that starts on line 13.

20 This is a sum statement. A sum statement is identified by the absence of an equal sign (=). It means: *Add the current value of AMOUNT to CAT_SUM.* The fact that CAT_SUM is named on the left side of a sum statement means that it will be given an initial value of zero and will not be reset to missing value (null) before each new observation is read. CAT_SUM accumulates the dollar total of all claim amounts for the current injury category and is reset to zero only at the start of a new category. (See line 16.)

21 Like line 20 this is a sum statement. Here, 1 is added to the variable CAT_CNT for each observation in the current injury category. This statement counts claims for each category. It is reset to zero at the start of a new category. (See line 17.)

> **Closer Look**
>
> Program 15 uses a SAS data set as input. **The DATA step can read external files as well as SAS data sets** (see Chapter 2 in this book). This means you can use the DATA step language to generate reports directly from external files. However, the BY statement and FIRST. and LAST. variables are available only when reading a SAS data set.

Program 15 Notes

Line

23 This IF statement tests to see if there are less than three lines remaining on the page. Why check for three lines here? If the link to the summary routine, SUM_CAT, is executed on line 32, the next statements could print up to three lines on the page: one detail line (as a result of the PUT statement on lines 28-30) plus the two summary lines (as a result of the LINK statement on line 32).

24 **PUT _PAGE_** moves the print position to a new page.

25 This LINK statement executes the header routine that begins at line 36. You need to print the header here because the preceding PUT statement moved the print position to the top of a new page.

26 The END statement ends the DO group that began on line 23.

28-30 This PUT statement writes a detail line—the information for each claim. The style used here is called *formatted PUT*. It consists of three parts:

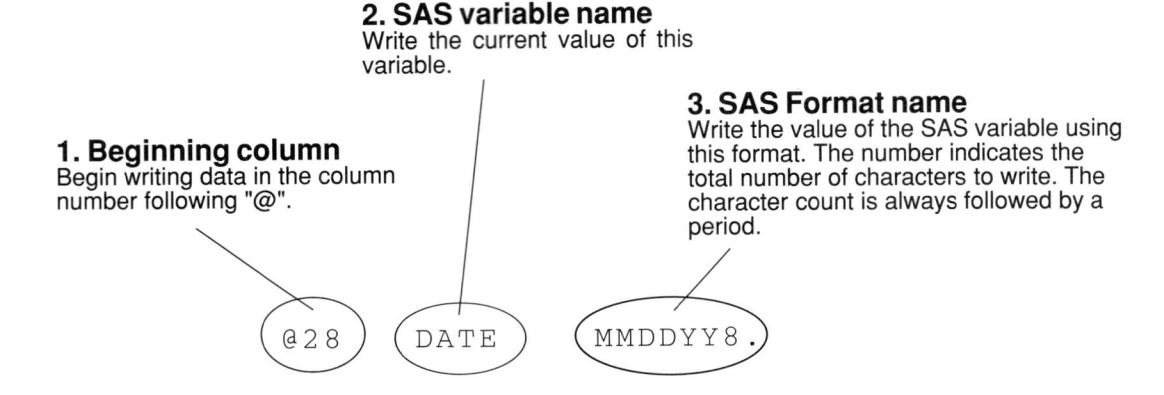

2. SAS variable name
Write the current value of this variable.

1. Beginning column
Begin writing data in the column number following "@".

3. SAS Format name
Write the value of the SAS variable using this format. The number indicates the total number of characters to write. The character count is always followed by a period.

(@28) (DATE) (MMDDYY8.)

AMOUNT is written in nine column spaces with two decimal places (9.2 format). DATE is written with the MMDDYY8. format. There is no format specification for REGION, so it is written in the default format for character variables.

32 The IF statement tests LAST.CATEGORY. When it has a value of 1, the current observation is the last in an injury category and the LINK statement is executed, causing a jump to the category summary information routine starting after the label on line 51.

34 This RETURN statement causes control to jump back to the top of the DATA step. The SET statement is executed again, and the next observation is read. Without the RETURN statement, the DATA step would "fall through" to the category header (CAT_HDR) statements for every observation. You want the category header routine executed only by the LINK statements on lines 15 and 25.

36 CAT_HDR is a statement label. It serves as the target for the LINK statements in lines 15 and 25. The underscore (_) is a legal SAS name character. Labels always end with a colon (:).

Program 15 Notes

Line 37-48 This PUT statement writes the injury category header. Like all SAS statements, it begins with a keyword (PUT) and ends with a semicolon (on line 48). Each slash (/) moves the print position to a new line. The two slashes on line 37 cause two blank lines to appear before any header text. The value of CATEGORY is printed at column 1, and the literal text "Amount" at column 17. The slash on line 40 causes the print position to move to a new line. The literal "Claims" is written at column 1 of the new line, "of Claim" at column 16, etc.

49 The RETURN statement causes control to return to the statement following the active LINK. When the LINK statement on lines 15 or 25 is executed, the DATA step jumps to the CAT_HDR label (line 36) and executes the statements on the following lines. This RETURN causes control to jump back to line 16 or 26–the line following the active LINK statement.

51 SUM_CAT is a statement label. It is the target of the LINK statement on line 32.

52 This PUT statement prints the summary information. The value of CAT_CNT, the count of claims for each injury category, is printed after the literal text "Count: ". Next, a line of equal signs is printed beginning in column 13. The slash moves the print position to the next line. CAT_SUM is printed at column 11 with the DOLLAR13.2 format (see End Notes). CAT_SUM holds the dollar total of claims for the current injury category as accumulated on line 20.

57 This RETURN statement acts like the one on line 49. It causes control to return to the statement following the active LINK statement. The active LINK statement is on line 32, so control returns to the statement on line 34.

59 The RUN statement ends and executes the DATA step.

Closer Look

The RETURN statement can have two meanings depending on whether or not a LINK statement is active. When a RETURN statement appears at the end of a group of statements executed as the result of a LINK statement, it returns control to the statement immediately following the calling LINK statement, as in lines 49 and 57 in Program 15. When a LINK statement is not active, the RETURN statement causes control to jump to the top of the DATA step, as in line 34.

✱ Results Custom Layout Claims Report

The resulting claim category report appears below and on the next page. You can cross check the headers and summation lines with corresponding statements in Program 15.

DATA step report writing gives you complete control over the layout of your report, but at a cost. DATA step programs can be complicated when there are multiple BY-group levels with summations and counts. When you use the DATA step you are responsible for all page formatting. Still, the DATA step language, with its built-in features such as the LL= option and BY-group processing, makes the SAS DATA step much easier to use than procedural languages like COBOL or BASIC.

Claims report, page 1

```
                        Claim Category Report
         INJURY         Amount
         Claims         of Claim      Date        Region
                        ========      ========    ======
                        19551.32      04/28/93    NORTH
                        19225.54      11/29/93    NORTH
                        18963.10      04/15/93    WEST
                        18264.37      06/08/93    SOUTH
                        16837.60      04/28/93    EAST
                        16679.13      04/07/93    SOUTH
                        16533.09      08/03/93    EAST
                        14454.19      01/04/93    NORTH
                        13966.44      11/14/93    WEST
                        10780.58      01/10/93    SOUTH
                        10681.64      01/09/93    NORTH
                         9060.88      04/24/93    SOUTH
                         7797.22      02/23/93    SOUTH
                         6447.74      03/19/93    EAST
                         5248.55      04/17/93    EAST
                         3743.85      02/26/93    SOUTH
         Count: 16      ===========
                        $208,235.24

         NON-INJURY     Amount
         Claims         of Claim      Date        Region
                        ========      ========    ======
                        19829.24      12/04/93    NORTH
                        19770.49      10/05/93    NORTH
                        19748.66      05/21/93    SOUTH
                        19511.68      09/27/93    EAST
                        18539.09      07/14/93    EAST
                        17991.21      12/24/93    SOUTH
                        16091.38      08/28/93    SOUTH
                        15822.52      07/28/93    WEST
                        15316.09      12/05/93    SOUTH
                        14135.71      09/08/93    NORTH
                        12661.89      06/29/93    SOUTH
                        11462.81      05/20/93    SOUTH
                        11434.47      05/11/93    SOUTH
                        11269.33      05/27/93    NORTH
                        11231.73      05/18/93    EAST
                        10137.39      08/18/93    EAST
```

Claims report, page 2

```
                           Claim Category Report
             NON-INJURY      Amount
             Claims         of Claim      Date        Region
                            ========    ========      ======
                            10066.23    03/04/93      WEST
                             9984.92    07/20/93      SOUTH
                             9971.64    10/11/93      WEST
                             9515.56    09/17/93      NORTH
                             9515.27    02/27/93      NORTH
                             9140.13    05/14/93      SOUTH
                             8375.73    01/12/93      NORTH
                             8191.58    07/03/93      NORTH
                             7846.33    05/03/93      SOUTH
                             7838.12    01/19/93      WEST
                             7830.25    03/01/93      WEST
                             7823.31    12/08/93      SOUTH
                             7422.62    08/28/93      WEST
                             7135.95    07/16/93      WEST
                             6643.65    09/29/93      WEST
                             6362.79    01/09/93      SOUTH
                             6281.82    09/08/93      SOUTH
                             5913.21    11/05/93      WEST
                             5814.65    06/14/93      EAST
                             5544.70    07/02/93      EAST
                             5350.47    12/19/93      SOUTH
                             5279.15    12/14/93      EAST
                             4628.15    08/11/93      NORTH
                             3785.98    04/21/93      EAST
                             3284.94    10/19/93      WEST
                             3179.99    07/01/93      SOUTH
                             3015.00    06/12/93      WEST
                             2917.64    03/15/93      EAST
                             2024.45    11/30/93      WEST
                             1573.88    07/11/93      NORTH
                             1468.71    06/16/93      WEST
                             1129.14    12/23/93      EAST
                             1026.76    06/10/93      WEST
             Count: 49      ===========
                            $440,836.41
```

Chapter 15 - Creating a Custom Report 159

Reporting Features, Statements, and Alternatives

Report writing features. DATA step report writing can be complicated, and Program 15 is no exception. One approach to the problem is to list your requirements and match those requirements with DATA step features. Report writing features can be found in a variety of DATA step statements and options. The following table can help you determine which DATA step statements and options you'll need to generate a custom report. The line numbers in parentheses refer to examples in Program 15.

Report Writing Requirement	SAS System Feature
start a report writing DATA step	DATA _NULL_ statement (line 5)
direct output to the standard print destination	FILE statement with PRINT filename (line 8)
group, rearrange, or sort data	PROC SORT step (lines 1-3)
print a title on each page	TITLE statement in or before the DATA step (line 7)
read data from a SAS data set	SET statement (line10)
handle data groups	PROC SORT (lines 1-3), BY statement following the SET statement (line 11), IF statements with FIRST.byvariable and LAST.byvariable (lines 13 and 32)
write a report line with text positioned at a specific column	PUT statement with formatted output: (@ column with SAS format) (lines 28-30, 37-48, and 52-56)
move the print position to the next line	PUT statement with "/" (lines 37, 40, 45 and 54)
accumulate BY-group totals	SUM statement (line 20), reset the accumulation variable to 0 when the BY group changes (line 16)
count occurrences within a BY group	SUM statement, add 1 to a counter variable for each observation (line 21), reset the counter for each new BY group (line 17)
determine the number of printable lines left on the page	LL= option in the FILE statement (line 8), test LL= variable with an IF statement (lines 14 and 23)
execute statements only when a new BY group begins	IF statement with FIRST.variable and LINK statement, use RETURN statement to resume processing (lines14-15, 34, 50)
execute statements only when a BY group ends	IF statement with LAST.variable and LINK statement (line 32), use RETURN statement to resume processing (line 57)
separate the main DATA step processing statements from linked subroutines	RETURN statement, returns control to the top of the DATA step (line 34)
label linked subroutines	statement label (lines 36 and 51)
terminate linked subroutines	RETURN statement, returns control to the statement following the active link (lines 49 and 57)

- **Page settings.** When generating reports, you should be aware of the PAGESIZE= (or PS=) and LINESIZE= (or LS=) SAS system options. The PAGESIZE= option determines how many lines are on a page. It is not directly related to the physical length of a piece of paper in your printer. You can set PS= to any value between 15 and 32,767, regardless of the size of the paper you are using. Normally, you do want the actual number of lines available on the paper to match PAGESIZE=.

  When *PAGESIZE* number of lines have been printed, the printer position moves to a new page and the TITLE statement text is printed along with any other new page items such as page numbers and date stamps. The DATA step uses the PAGESIZE= setting to determine the value of the LL= variable as follows:

  value of LL= variable = PAGESIZE – number of lines already written to the current page

  The LINESIZE= option determines how many column positions are on a line. You can set LINESIZE= to any number between 64 and 256. Like PAGESIZE=, LINESIZE= does not necessarily correspond to a physical page width, but normally it should be set to coincide to the number of columns on the paper on which your report will be printed. When data are centered on a page, they are centered between column 1 and the LINESIZE= setting.

  When working with the SAS Display Manager System it is common to set LINESIZE= and PAGESIZE= to correspond to the number of columns and lines on your display. Be sure to reset these options to correspond to the actual paper size before running SAS steps that generate output intended for hardcopy printing.

  Finally, be aware that the SAS System assumes that attached printers are fixed pitch; that is, all characters take up the same amount of space. If your printer is set up to use a proportionally spaced font, the LINESIZE= setting will not determine the actual print width, only the total number of characters. With proportional fonts precise column alignment can be difficult.

  For details, see "SAS System Options," in Chapter 3, "Components of the SAS System," *SAS Language: Reference, Version 6, First Edition.*

- **Dollar signs.** The DOLLAR13.2 format is used on line 56 of Program 15. The DOLLAR format displays numbers with dollar signs and commas. To use the format you specify a name, "DOLLAR", and a length, "13.". The length is always followed by a period. Optionally, you can specify the number of decimal places to show after the period. The DOLLAR13.2 format shows two decimal places. The total number of characters displayed is still 13, so the largest number that can be displayed without truncation is "$9,999,999.99". Each character, including dollar signs and commas, counts in the length, not just digits. When you use a SAS format to display values that have a decimal component but do not specify a number of decimal places to the right of the period, values are rounded to the nearest whole number.

📝 **Report writing alternatives.** There are several reporting procedures available in the SAS System. These procedures allow you to create reports with very few statements and without worrying about issues such as centering data on the page, page breaks, printing group and overall totals in the right place, etc. The trade off in using these procedures versus a DATA step program is between ease of use and flexibility. SAS procedures offer several formatting and presentation controls, but you are limited to a fairly narrow range of standard layouts. The layout required for the claim report cannot be exactly duplicated with a SAS procedure. However, you can come close with PROC PRINT as follows:

```
PROC SORT DATA=AUTODIV.CLAIMS;
   BY CATEGORY DESCENDING AMOUNT;
RUN;

PROC PRINT DATA=AUTODIV.CLAIMS N SPLIT='/';
   TITLE 'Claim Category Report';
   LABEL AMOUNT='Amount /of Claim' DATE='Date'
         REGION='Region';
   FORMAT DATE MMDDYY8. AMOUNT DOLLAR13.2;
   BY CATEGORY;
   ID CATEGORY;
   VAR AMOUNT DATE REGION;
   SUM AMOUNT;
RUN;
```

The first page of the resulting report is shown on the following page. If you can live with the layout, a SAS procedure step will almost always be easier to write and maintain than a DATA step program. See Chapter 27, "The PRINT Procedure," *SAS Procedures Guide, Version 6, Third Edition,* for details on PROC PRINT features and options.

Another alternative to DATA step report writing is PROC REPORT. PROC REPORT is part of base SAS software. You can use it interactively to set up and run report definitions. For details see the *SAS Guide to the REPORT Procedure: Usage and Reference, Version 6, First Edition.*

162 Part 3 - Presenting Your Data

Claims report from PROC PRINT, first page

```
                   Claim Category Report

                    Amount
       CATEGORY    of Claim          Date      Region

       INJURY      $19,551.32      04/28/93    NORTH
                   $19,225.54      11/29/93    NORTH
                   $18,963.10      04/15/93    WEST
                   $18,264.37      06/08/93    SOUTH
                   $16,837.60      04/28/93    EAST
                   $16,679.13      04/07/93    SOUTH
                   $16,533.09      08/03/93    EAST
                   $14,454.19      01/04/93    NORTH
                   $13,966.44      11/14/93    WEST
                   $10,780.58      01/10/93    SOUTH
                   $10,681.64      01/09/93    NORTH
                    $9,060.88      04/24/93    SOUTH
                    $7,797.22      02/23/93    SOUTH
                    $6,447.74      03/19/93    EAST
                    $5,248.55      04/17/93    EAST
                    $3,743.85      02/26/93    SOUTH
       ----------  -------------
       INJURY      $208,235.24

                       N = 16


                    Amount
       CATEGORY    of Claim          Date      Region

       NON-INJURY  $19,829.24      12/04/93    NORTH
                   $19,770.49      10/05/93    NORTH
                   $19,748.66      05/21/93    SOUTH
                   $19,511.68      09/27/93    EAST
                   $18,539.09      07/14/93    EAST
                   $17,991.21      12/24/93    SOUTH
                   $16,091.38      08/28/93    SOUTH
                   $15,822.52      07/28/93    WEST
                   $15,316.09      12/05/93    SOUTH
                   $14,135.71      09/08/93    NORTH
                   $12,661.89      06/29/93    SOUTH
                   $11,462.81      05/20/93    SOUTH
                   $11,434.47      05/11/93    SOUTH
                   $11,269.33      05/27/93    NORTH
                   $11,231.73      05/18/93    EAST
                   $10,137.39      08/18/93    EAST
                   $10,066.23      03/04/93    WEST
                    $9,984.92      07/20/93    SOUTH
```

Quick Summary

Creating a Custom Report

1. Make sure you know the name and have access to the SAS data set where your data are stored. This may require a LIBNAME statement. (See Chapter 2 in this book for a LIBNAME example.)

2. Use a DATA step. Read observations with the SET statement.

3. To handle data in groups, use the BY statement to create FIRST.*variable* and LAST.*variable* temporary variables.

4. Use the FILE statement with the PRINT filename to send output to the standard SAS print output target. Use the FILE statement option LL=*variable-name* to determine the number of lines remaining on a page and handle page breaks.

5. Use the PUT statement to write report lines.

6. Use LINK and RETURN statements to execute and return from header and data group summation routines.

DATA Step Reporting Features

BY groups, FIRST.*variable* and LAST.*variable*
"BY-Group Processing" in Chapter 4, "Rules of the SAS Language," *SAS Language: Reference, Version 6, First Edition*

DATA step overview
Chapter 2, "The DATA Step," *SAS Language: Reference*

DATA step report writing overview and examples
"Writing a Report" in Chapter 29, "Writing Output," *SAS Language and Procedures: Usage, Version 6, First Edition*

FILE statement and options
"FILE" in Chapter 9, "SAS Language Statements," *SAS Language: Reference*

IF statement and DO groups
"Performing More than One Action in One IF-THEN Statement" in Chapter 12, "Finding Shortcuts in Programming," *SAS Language and Procedures: Usage;*
"IF-THEN/ELSE" and "DO" in Chapter 9, "SAS Language Statements," *SAS Language: Reference*

LINK and RETURN statements
"LINK" and "RETURN" in Chapter 9, "SAS Language Statements," *SAS Language: Reference*

LL= option
"FILE" in Chapter 9, "SAS Language Statements," *SAS Language: Reference*

PUT statement, _PAGE_
"Writing Simple Text" in Chapter 29, "Writing Output," *SAS Language and Procedures: Usage;*
"PUT" in Chapter 9, "SAS Language Statements," *SAS Language: Reference*

SAS formats, date
"Displaying Dates" in Chapter 13, "Working with Dates in the SAS System," *SAS Language and Procedures: Usage*

SAS formats, general
Chapter 14, "SAS Formats," *SAS Language: Reference*

SET statement
"SET" in Chapter 9, "SAS Language Statements," *SAS Language: Reference*

sum statement
"Getting a Total for Each BY Group" in Chapter 11, "Using More than One Observation in a Calculation," *SAS Language and Procedures: Usage;*
"Sum" in Chapter 9, "SAS Language Statements," *SAS Language: Reference*

Chapter 16
Creating Bar Charts

Learn How to...

- Create a vertical bar chart
- Select a graphics output device
- Select bar patterns and text fonts

Using These SAS System Features...

- GCHART procedure
- VBAR statement and options
- GOPTIONS statement
- PATTERN statement

? Problem Create an Hourly Utilization Chart

A data processing center records the percentage of computer system capacity used each hour of the day. The information is stored in SAS data set PERFORM.CAPUTIL, which appears below. The variable HOUR is the time of day on a 24-hour clock. UTILIZE is the percentage of total capacity used that hour.

As part of a capacity planning project, you need to create a bar chart showing hourly utilization. The horizontal axis should be the hour of the day, and the height of each bar should represent the utilization percentage.

SAS data set PERFORM.CAPUTIL

OBS	HOUR	UTILIZE
1	0	20
2	1	24
3	2	25
4	3	29
5	4	27
6	5	37
7	6	42
8	7	48
9	8	67
10	9	93
11	10	96
12	11	95
13	12	77
14	13	93
15	14	92
16	15	80
17	16	81
18	17	72
19	18	38
20	19	31
21	20	28
22	21	28
23	22	23
24	23	21

GCHART Procedure and VBAR Statement

Solution

Program 16 uses SAS/GRAPH software's GCHART procedure, plus global statements GOPTIONS, PATTERN, and AXIS, to generate the utilization chart. PROC GCHART and associated global statements give you extensive control over both the appearance and content of your charts.

Program 16

```
1   GOPTIONS DEVICE=HP7440 ROTATE=LANDSCAPE;
2
3   PATTERN VALUE=SOLID;
4
5   AXIS1 LABEL=(HEIGHT=3 PCT FONT=SWISS 'Percent');
6   AXIS2 LABEL=(HEIGHT=3 PCT FONT=SWISS 'Hour of the Day');
7
8   PROC GCHART DATA=PERFORM.CAPUTIL;
9      TITLE FONT=SWISS HEIGHT=5 PCT 'System Utilization';
10     VBAR HOUR /
11        SUMVAR=UTILIZE
12        DISCRETE
13        RAXIS=AXIS1
14        MAXIS=AXIS2
15     ;
16
17  RUN;
```

Closer Look

When running PROC GCHART in the SAS Display Manager System or interactive line mode, be aware that **the RUN statement does not terminate the procedure.** Once PROC GCHART is started you can submit any number of VBAR statements, each followed by a RUN statement. The procedure executes your statements but remains active, waiting for more. The procedure is only terminated by the submission of another PROC or DATA step, or the QUIT statement. You can tell if a procedure is still active by checking the title bar of the PROGRAM EDITOR window. See "Using RUN Groups" in Chapter 2, "Running SAS/GRAPH Programs," in *SAS/GRAPH Software: Reference, Version 6, First Edition, Volume 1* for more information.

Program 16 Notes

Line

1 The GOPTIONS statement allows you to set several options that affect graphics output. The DEVICE= option names the output device for the bar chart, here a Hewlett-Packard 7440 plotter (see Closer Look below). The ROTATE= option specifies page orientation: LANDSCAPE or PORTRAIT. Portrait orientation means the long side of the page is vertical; with landscape, the long side of the page is horizontal.

3 When used with PROC GCHART, the PATTERN statement options affect the appearance of the chart bars. Here, the VALUE= option specifies a solid fill (SOLID). See End Notes for more on the PATTERN statement.

5 The AXIS1 statement creates a set of axis specifications. These specifications will be assigned to the chart's vertical axis. (See line 13 on the next page.) You can define up to 99 sets of axis definitions: AXIS1 through AXIS99.

There are several options available in an AXIS statement. The LABEL= option controls the text used to label an axis. Between the parenthesis following the equal sign is a set of text description parameters plus a text literal enclosed in quotes. The text description settings are HEIGHT=3 PCT, and FONT=SWISS. "Percent" will be used as the axis label. The height of the characters will be 3% of the total graphics area on the page and output will appear in the SWISS font.

6 The AXIS2 statement creates a second set of axis specifications that will be assigned to the chart's horizontal axis. (See line 14 on the next page.) The axis label text will be "Hour of the Day". The height of this text will be 3% of the graphics area, and the font will be SWISS.

GOPTIONS, PATTERN, and AXIS are global statements. They are independent of any one procedure. They could be used by any procedure that follows them in your program. Once global settings are specified they are in effect for all subsequent procedures in your SAS session, until explicitly changed.

SAS/GRAPH software offers extensive support for graphics output devices. Graphics output is implemented via device drivers. Drivers are supplied for dozens of target graphics displays, printers, and output files. You can even modify many parameters of individual drivers, effectively creating your own customized device drivers. When running in display manager, you can check or change the DEVICE= option from the OPTIONS window. You can get a list of drivers by submitting the following step:

```
PROC GDEVICE;
RUN;
```

For an example see Chapter 1, "Producing Graphics Output in the SAS System" in *SAS/GRAPH Software: Introduction, Version 6, First Edition*. For details on listing and modifying SAS graphics drivers see Chapter 25, "The GDEVICE Procedure" in *SAS/GRAPH Software: Reference, Version 6, First Edition, Volume 2*.

Program16 Notes

Line   **8**   The PROC statement starts the GCHART procedure and names PERFORM.CAPUTIL as the input SAS data set.

         9   The TITLE statement defines a title for the PROC GCHART output. The FONT= option specifies the SWISS typeface, and the HEIGHT= option specifies the size of the text as 5% of the available graphics area. The quoted text will appear at the top of each page.

       10   The VBAR statement begins on this line and ends on line 15 with the semicolon. VBAR names the type of chart you want: a vertical bar chart. PROC GCHART can also generate horizontal bar (HBAR) and other charts.

               HOUR names the charting variable. You can interpret VBAR *variable* as: *Show a bar for each value of variable-name.* Here it means: *Show a bar for each value of HOUR.*

               The "/" separates the charting variable name from VBAR options.

       11   The SUMVAR= option determines what is represented by the height of each bar. SUMVAR=UTILIZE means: *Make the height of each bar represent the total of UTILIZE.* (See Closer Look below.)

       12   The DISCRETE option means: *Show a bar for each unique value of the charting variable HOUR, no matter how many there are.* (See End Notes for more on the DISCRETE option.)

       13   The RAXIS= option assigns the AXIS1 settings to the response axis. The response axis applies to the SUMVAR variable (UTILIZE). For VBAR charts the response axis is the vertical axis.

       14   The MAXIS= option assigns the AXIS2 settings to the midpoint axis. The midpoint axis applies to the charting variable (HOUR). For VBAR charts the midpoint axis is the horizontal axis.

       15   The semicolon ends the VBAR statement.

       17   The RUN statement executes the GCHART procedure.

Closer Look

In Program 16, the **SUMVAR= option** tells PROC GCHART to sum the variable UTILIZE for each value of the charting variable (HOUR). In the PERFORM.CAPUTIL data set, there is only one observation for each HOUR. If there were five observations for HOUR, there would still be just one bar. The height of that bar would represent the sum of all five UTILIZE values for each value of HOUR.

Chapter 16 - Creating Bar Charts   **171**

✳ Results | Capacity Utilization Chart

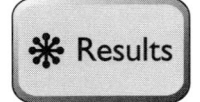

The capacity utilization chart generated by Program 16 appears below. As required, the bars show the value of UTILIZE for each value of HOUR. The AXIS1 statement (line 5) and RAXIS= option (line 13) cause the vertical axis to be labeled as "Percent" instead of "UTILIZE" and determine the size and typeface of the label. The AXIS2 statement and MAXIS option determine the text and appearance of the horizontal axis label. The title text is controlled by the TITLE statement on line 9.

Utilization chart

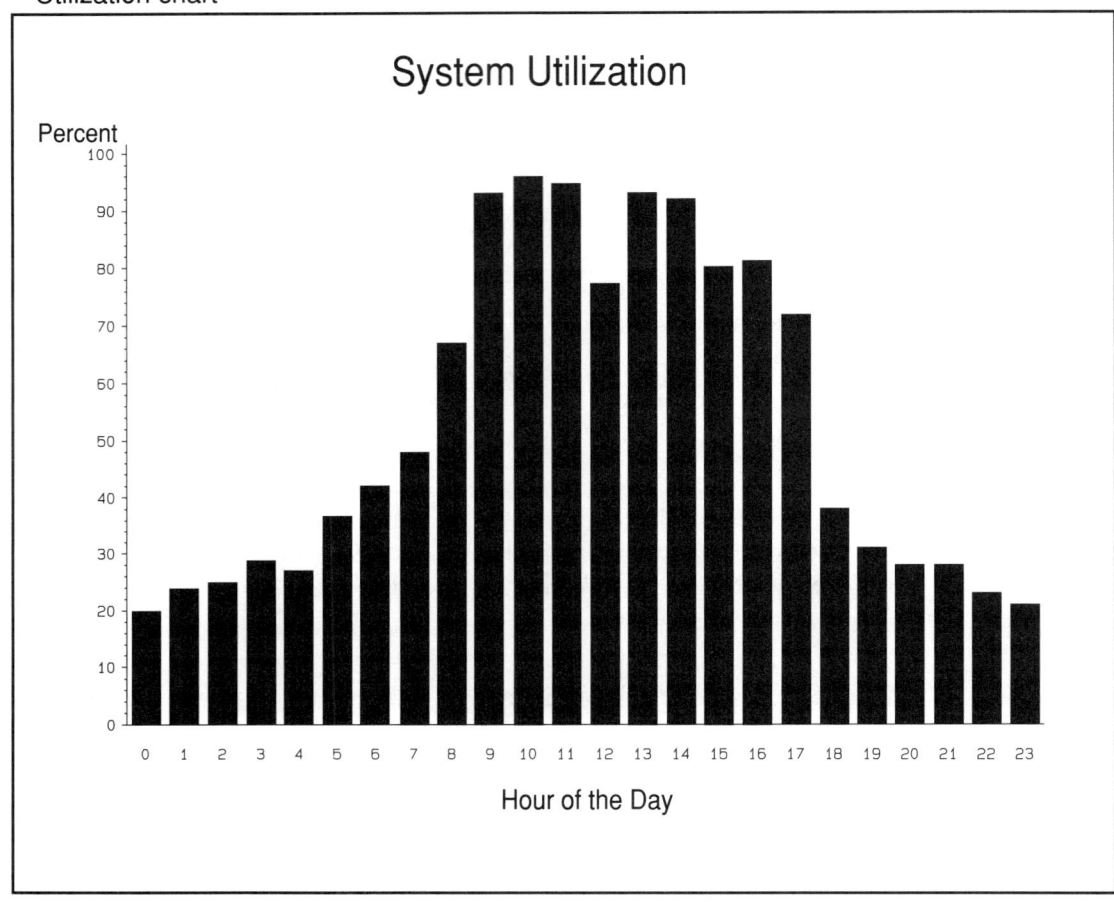

System Utilization — vertical bar chart showing Percent (0–100) on the y-axis versus Hour of the Day (0–23) on the x-axis. Approximate values: hour 0: 20, 1: 24, 2: 25, 3: 29, 4: 27, 5: 37, 6: 42, 7: 48, 8: 67, 9: 93, 10: 96, 11: 95, 12: 77, 13: 93, 14: 92, 15: 80, 16: 81, 17: 72, 18: 38, 19: 31, 20: 28, 21: 28, 22: 23, 23: 21.

🔍 Closer Look

Unless otherwise specified, the horizontal **x-axis values are displayed in ascending order** (here, 0 to 23) regardless of the input data set order. You can alter the sequence of bars with the MIDPOINTS= option in the VBAR statement or the ORDER= option in the AXIS statement. See "VBAR Statement" in Chapter 23, "The GCHART Procedure," *SAS/GRAPH Software: Reference, Version 6, First Edition, Volume 2,* and "AXIS Statement Options" in Chapter 9, "The AXIS Statement," *SAS/GRAPH Software: Reference, Version 6, First Edition, Volume 1.*

End Notes: More on Creating Charts

Nongraphic graphics. The solutions in this chapter and in Chapters 17, 18, and 19 assume that you have access to SAS/GRAPH software. SAS/GRAPH software allows you to specify colors, patterns, text fonts and sizes, as well as send output to graphics devices. You can also create charts with base SAS software using the CHART procedure. The control statements for PROC CHART are similar to those for PROC GCHART. The difference is that PROC CHART produces character output only. This is not as attractive as the graphical output of PROC GCHART, but it can be displayed on any character-only terminal and printed on any standard character printer. SAS/GRAPH global statements such as GOPTIONS, PATTERN, and AXIS are not supported by PROC CHART. Following is the PROC CHART version of the hourly utilization bar chart:

```
PROC CHART DATA=PERFORM.CAPUTIL;
    TITLE 'System Utilization Percentage';
    VBAR HOUR /
       SUMVAR=UTILIZE
       DISCRETE
    ;
RUN;
```

The resulting chart appears below:

PROC CHART version of the utilization chart

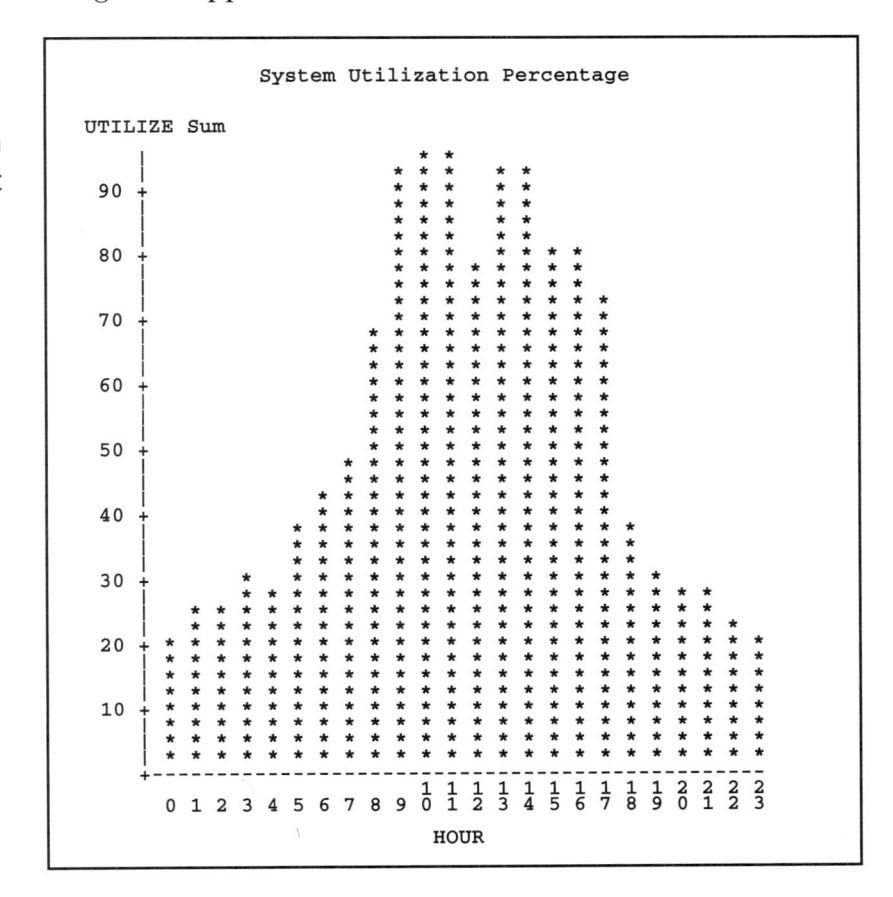

```
                    System Utilization Percentage
   UTILIZE Sum
                              * *
                            * * *     * *
    90 +                    * * *     * *
                            * * *     * *
                            * * *     * *
                            * * *     * *
    80 +                    * * * * * * *
                            * * * * * * * *
                            * * * * * * * *
    70 +                  * * * * * * * * *
                          * * * * * * * * *
                          * * * * * * * * *
    60 +                  * * * * * * * * *
                          * * * * * * * * *
                          * * * * * * * * *
    50 +                * * * * * * * * * *
                        * * * * * * * * * *
                        * * * * * * * * * *
    40 +              * * * * * * * * * * *   *
                      * * * * * * * * * * * * *
                      * * * * * * * * * * * * *
    30 +          *   * * * * * * * * * * * * *
               * * * * * * * * * * * * * * * *
               * * * * * * * * * * * * * * * *
    20 + * * * * * * * * * * * * * * * * * * * *
         * * * * * * * * * * * * * * * * * * * *
         * * * * * * * * * * * * * * * * * * * *
    10 + * * * * * * * * * * * * * * * * * * * *
         * * * * * * * * * * * * * * * * * * * *
         * * * * * * * * * * * * * * * * * * * *
       +-----------------------------------------
                           1 1 1 1 1 1 1 1 1 1 2 2 2 2
       0 1 2 3 4 5 6 7 8 9 0 1 2 3 4 5 6 7 8 9 0 1 2 3
                              HOUR
```

Charts without SUMVAR. The height of the bars in your chart can represent several different quantities. Without the SUMVAR= option each bar will represent frequency, that is, the count of each value of HOUR in the data set. The count of each unique value of HOUR in PERFORM.CAPUTIL is 1. There is one observation for each of the 24 hours. To illustrate, SUMVAR= is absent in the following program:

```
GOPTIONS DEVICE=HP7440 ROTATE=LANDSCAPE;
PATTERN VALUE=SOLID;
PROC GCHART DATA=PERFORM.CAPUTIL;
   TITLE FONT=SWISS
         HEIGHT=5 PCT 'System Utilization Percentage';
   VBAR HOUR /
      DISCRETE
   ;
RUN;
```

The resulting chart appears below:

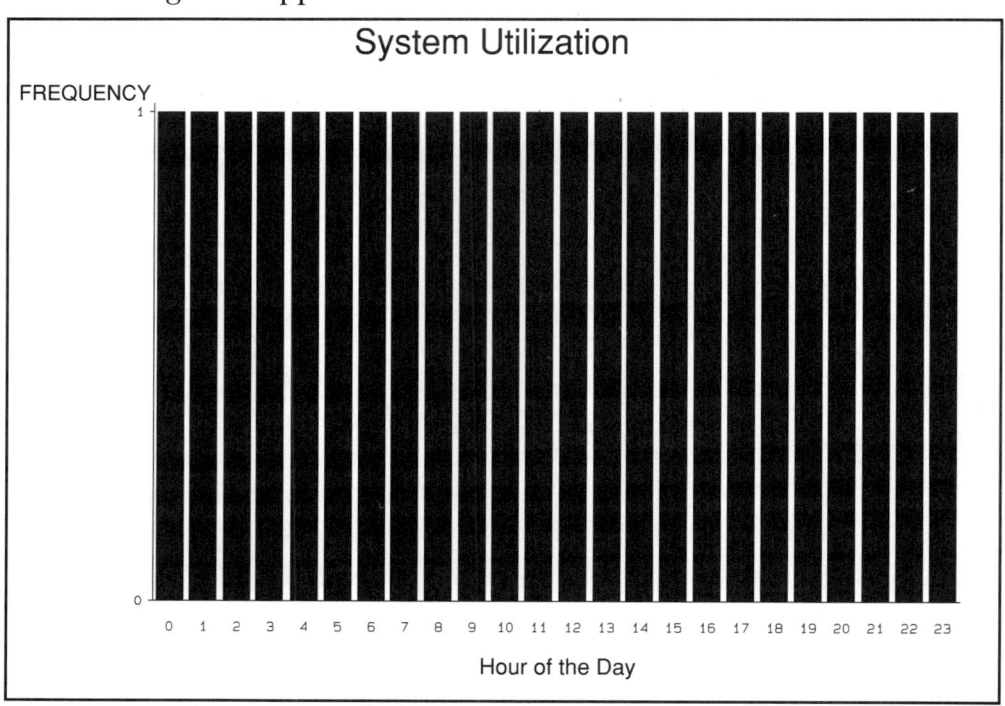

Utilization chart with SUMVAR= removed

Each bar has a height of 1 because each unique value of HOUR in the data set PERFORM.CAPUTIL appears just once.

174 Part 3 - Presenting Your Data

The DISCRETE option. When a charting variable is numeric, as is HOUR in the PERFORM.CAPUTIL data set, PROC GCHART calculates a number of levels or ranges of values by default instead of displaying a bar for each unique value of the numeric variable. The DISCRETE option forces the procedure to abandon this default and treat each value individually. Perhaps the easiest way to explain the effect of the DISCRETE option is to show what happens without it. The DISCRETE option is removed from the following program. Also note that the AXIS1 statement does not specify "Percent" as the label text as in Program 16. The vertical (RAXIS) label defaults to the SUMVAR variable name (UTILIZE) plus the default bar type (SUM).

```
GOPTIONS DEVICE=HP7440 ROTATE=LANDSCAPE;
PATTERN VALUE=SOLID;
AXIS1 LABEL=(HEIGHT=3 PCT FONT=SWISS);
AXIS2 LABEL=(HEIGHT=3 PCT FONT=SWISS 'Hour of the Day');

PROC GCHART DATA=PERFORM.CAPUTIL;
   TITLE FONT=SWISS HEIGHT=5 PCT 'System Utilization';
   VBAR HOUR /
      SUMVAR=UTILIZE
      RAXIS=AXIS1
      MAXIS=AXIS2
   ;
RUN;
```

The resulting chart output without the DISCRETE option appears below:

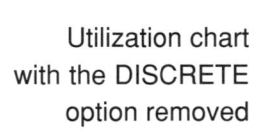

Utilization chart with the DISCRETE option removed

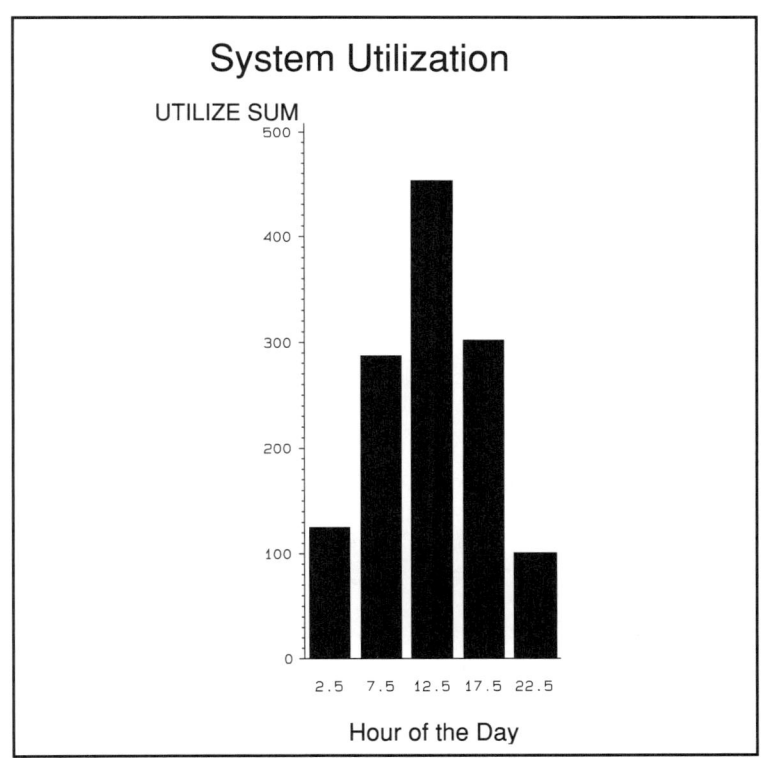

Without the DISCRETE option, PROC GCHART collapses the range of HOUR values into five ranges and generates a bar at the midpoint of each range.

Chapter 16 - Creating Bar Charts **175**

📝 **VBAR options.** The VBAR option TYPE= is not used in Program 16, but it is important to understand how this option can affect your charts. In Program 16, TYPE= is allowed to default to TYPE=SUM. You can use the TYPE= option to generate bars that represent mean (average), frequency, cumulative frequency, and frequency percentages. Consider the following statements:

```
VBAR HOURS / SUMVAR=UTILIZE TYPE=MEAN DISCRETE;
```

This statement creates a bar for each value of HOURS. The height of each bar represents the **mean** of UTILIZE.

```
VBAR HOURS / TYPE=PERCENT DISCRETE;
```

This statement creates a bar for each value of HOURS. The height of each bar represents the **percent** of all observations accounted for by each value of HOUR.

The various combinations of the SUMVAR= and TYPE= options can be a bit confusing. The following table lists the type of chart generated for each combination:

	When used with SUMVAR=	When used without SUMVAR=
TYPE=SUM *	sum of the SUMVAR= variable for each value of the charting variable	not valid
TYPE=MEAN	mean (average) of the SUMVAR= variable for each value of the charting variable	not valid
TYPE=FREQ **	not valid	number of observations for each unique value of the charting variable
TYPE=CFREQ	not valid	number of observations for each value of the charting variable, plus all preceding values
TYPE=PERCENT	not valid	percent of all observations accounted for by observations for each unique value of the charting variable
TYPE=CPERCENT	not valid	percent of all observations accounted for by observations for each unique value of the charting variable, plus all preceding values

* default when SUMVAR= option is used   ** default when SUMVAR= option is not used

Controlling appearance. The PATTERN, AXIS, and TITLE statements give you considerable control over the look of your chart. Two parameters you can specify in the PATTERN statement are color and fill pattern. The COLOR= option is only meaningful on a color or grey-scale output device. The fill pattern is named in the VALUE= option. Fills can be a solid color (VALUE=SOLID, as in Program 16), empty (VALUE=EMPTY), or one of several other patterns. The X3 hatch pattern (VALUE=X3) appears below. See Chapter 15, "The PATTERN Statement" in *SAS/GRAPH Software: Reference, Version 6, First Edition, Volume 1* for details and pattern examples.

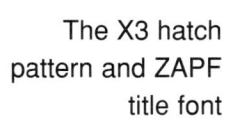

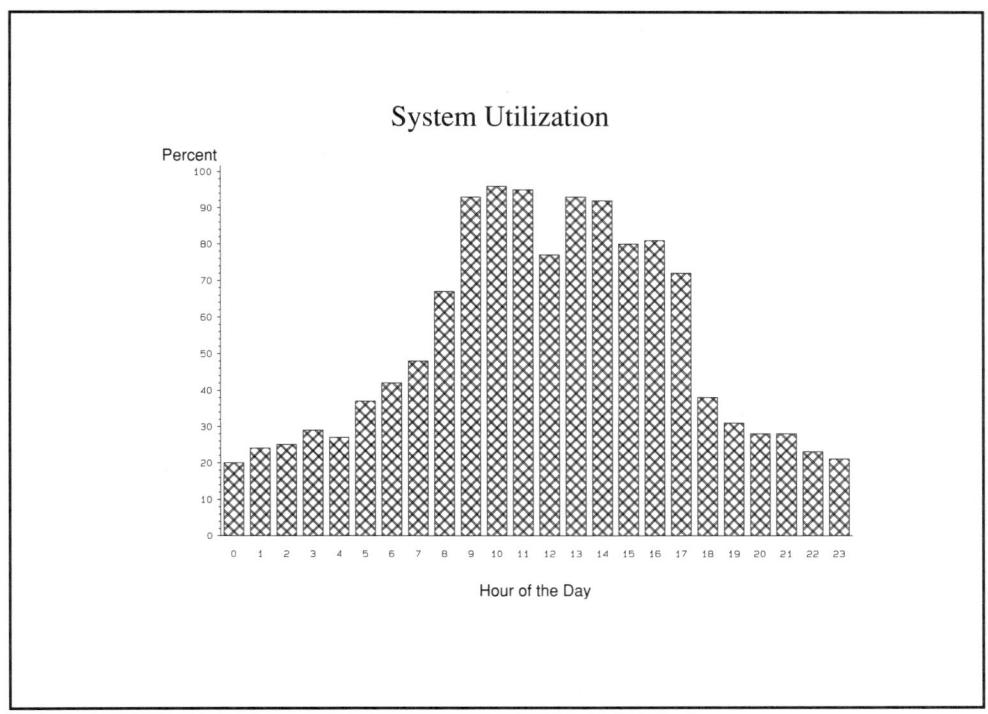

The X3 hatch pattern and ZAPF title font

Several TITLE and AXIS statement options are available with SAS/GRAPH procedures. These options allow you to specify a variety of appearance settings, including color and rotation. Program 16 uses FONT= and HEIGHT=. Several fonts are available. The ZAPF font (FONT=ZAPF) appears in the title text above. The HEIGHT= option affects the size of the text. The meaning of the number to the right of the equal sign in the HEIGHT= option can be centimeters, inches, character cells, or a percentage of the graphics area (PCT) as is Program 16. The default is character cells. See Chapter 6, "SAS/GRAPH Fonts" in *SAS/GRAPH Software: Reference, Volume 1* for font samples and usage details. For more on HEIGHT= see Chapter 17, "The TITLE Statement," and "How Graphics Are Drawn" in Chapter 2, "Running SAS/GRAPH Programs," in *SAS/GRAPH Software: Reference, Volume 1*.

Creating Bar Charts

1. Make sure you know the name and have access to the SAS data set with which you want to work. This may require a LIBNAME statement. (See Chapter 2 in this book for a LIBNAME example.)

2. Make sure you know the name of the driver for the graphics device where your chart will be printed or displayed. Name the graphics device in a GOPTIONS statement using the DEVICE= option. Use the GDEVICE procedure to list drivers.

3. Use PROC GCHART. Name the variable you want to chart in the VBAR statement. Values of this variable appear on the horizontal axis.

4. Use AXIS statements and RAXIS= and MAXIS= options to override default axis labeling.

5. With no VBAR options the height of each bar represents frequency. With the SUMVAR=*variable* option and no TYPE= option, each bar is the sum of *variable*. The DISCRETE option forces the display of a bar for each unique value of a numeric charting variable. The VBAR options list must be preceded by a "/".

PROC GCHART, Graphics Options

AXIS statement
Chapter 9, "The AXIS Statement," *SAS/GRAPH Software: Reference, Version 6, First Edition , Volume 1*

device drivers (graphics output devices)
Chapter 4, "Device Drivers," *SAS/GRAPH Software: Reference, Volume 1*

more bar chart examples
Part 2, "Bar Charts," *SAS/GRAPH Software: Usage, Version 6, First Edition*

PATTERN statement
"Specifying How Patterns Are Assigned" in Chapter 5, "Producing Horizontal Bar Charts," *SAS/GRAPH Software: Usage;*
Chapter 15, "The PATTERN Statement," *SAS/GRAPH Software: Reference, Volume 1*

PROC GCHART general reference
Chapter 23, "The GCHART Procedure," *SAS/GRAPH Software: Reference, Version 6, First Edition, Volume 2*

SAS graphics options (GOPTIONS)
Chapter 5, "Graphics Options and Device Parameters Dictionary," *SAS/GRAPH Software: Reference, Volume 1*

TYPE= option
"Selecting the Statistic to Chart" in Chapter 6, "Producing Vertical Bar Charts," *SAS/GRAPH Software: Usage;*
"VBAR Statement," in Chapter 23, "The GCHART Procedure," *SAS/GRAPH Software: Reference, Volume 2*

VBAR statement and options
Chapter 6, "Producing Vertical Bar Charts," *SAS/GRAPH Software: Usage;*
"VBAR Statement" in Chapter 23, "The GCHART Procedure," *SAS/GRAPH Software: Reference, Volume 2*

Chapter 17
Creating Stacked Bar Charts

Learn How to...

- Create a stacked bar chart
- Select a graphics output device
- Control chart text and bar fill patterns
- Organize data for charting

Using These SAS System Features...

- DATA step
- GCHART procedure
- SUBGROUP= option
- GOPTIONS statement
- PATTERN statement
- AXIS statement
- LEGEND statement

? Problem — Create a Utility Expense Chart

Each month, gas and electric utility costs are recorded in SAS data set EXPENSES.UTILITY. The data set has three variables: MONTH, a character variable holding the month of the bill (JAN to DEC); GAS, a numeric variable holding the natural gas portion of the bill; and ELEC, a numeric variable holding the electricity portion. EXPENSES.UTILITY appears below.

To track utility expenses you need to create a bar chart showing monthly utility costs. There should be one bar for each month showing the total utility cost. Each bar should be divided into electric and gas cost segments.

SAS data set EXPENSES.UTILITY

OBS	MONTH	GAS	ELEC
1	JAN	122.34	56.76
2	FEB	114.08	55.23
3	MAR	98.25	49.61
4	APR	77.67	43.98
5	MAY	48.01	40.77
6	JUN	39.80	39.07
7	JUL	37.91	39.82
8	AUG	38.47	39.74
9	SEP	41.35	40.27
10	OCT	68.50	45.34
11	NOV	97.72	48.99
12	DEC	108.83	53.70

GCHART Procedure and VBAR Statement

Solution

SAS/GRAPH software's GCHART procedure can easily generate the stacked bar chart required to show utility costs with gas and electric components. This type of chart is generated using the GCHART VBAR statement with the SUBGROUP= option.

When you use the SUBGROUP= option, the input data set must be structured properly. Program 17 consists of two SAS steps. The first, a DATA step, restructures the EXPENSES.UTILITY data set into a temporary data set called UTLTEMP. This data set has two observations for each month: one holding the gas expense information and one for electricity. Next, the GCHART procedure step uses the UTLTEMP data set as input to create a chart. The overall process is shown below. (The complete UTLTEMP data set is shown in End Notes.)

Restructuring the charting data for a stacked bar chart

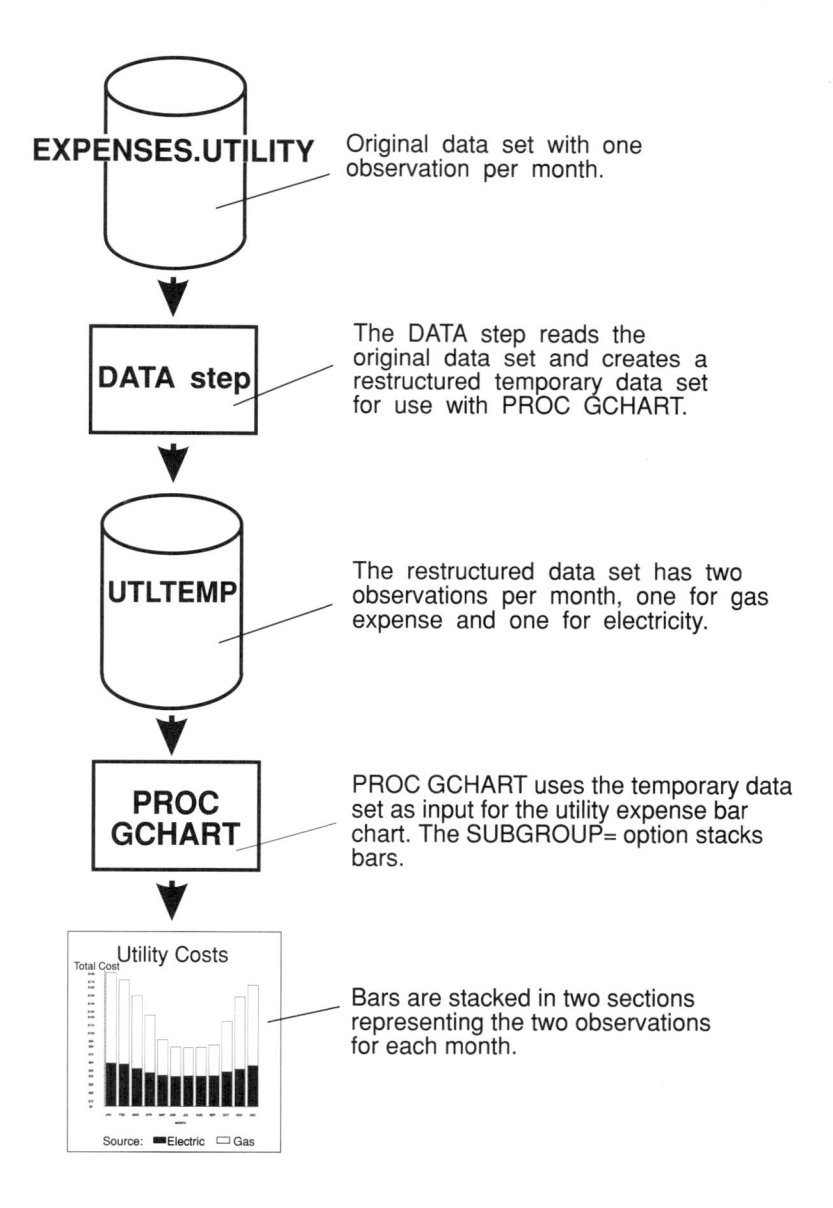

EXPENSES.UTILITY — Original data set with one observation per month.

DATA step — The DATA step reads the original data set and creates a restructured temporary data set for use with PROC GCHART.

UTLTEMP — The restructured data set has two observations per month, one for gas expense and one for electricity.

PROC GCHART — PROC GCHART uses the temporary data set as input for the utility expense bar chart. The SUBGROUP= option stacks bars.

Utility Costs — Bars are stacked in two sections representing the two observations for each month.

Chapter 17 - Creating Stacked Bar Charts 183

Program 17

```
 1  DATA UTLTEMP;
 2     KEEP MONTH TYPE AMOUNT;
 3     SET EXPENSES.UTILITY;
 4
 5     TYPE = 'E';
 6     AMOUNT = ELEC;
 7     OUTPUT;
 8
 9     TYPE = 'G';
10     AMOUNT = GAS;
11     OUTPUT;
12  RUN;
13
14  GOPTIONS DEVICE=HP7440 ROTATE=LANDSCAPE;
15
16  PATTERN1 VALUE=SOLID COLOR=BLACK;
17  PATTERN2 VALUE=EMPTY COLOR=BLACK;
18
19  AXIS1 LABEL=(HEIGHT=5 PCT FONT=SWISS 'Total Cost')
20        VALUE=(HEIGHT=4 PCT FONT=SWISS);
21  AXIS2 LABEL=(HEIGHT=5 PCT FONT=SWISS 'Month')
22        VALUE=(HEIGHT=2.5 PCT FONT=SWISS);
23
24  LEGEND1 LABEL=(HEIGHT=5 PCT FONT=SWISS 'Source:')
25          VALUE=(HEIGHT=5 PCT FONT=SWISS 'Electric'
26                 HEIGHT=5 PCT FONT=SWISS 'Gas');
27
28  PROC GCHART DATA=UTLTEMP;
29
30     TITLE HEIGHT=8 PCT FONT=SWISS 'Utility Costs';
31     FORMAT AMOUNT DOLLAR5.;
32
33     VBAR MONTH /
34        SUBGROUP=TYPE
35        SUMVAR=AMOUNT
36        RAXIS=AXIS1
37        MAXIS=AXIS2
38        LEGEND=LEGEND1
39        MIDPOINTS = 'JAN' 'FEB' 'MAR' 'APR' 'MAY' 'JUN'
40                    'JUL' 'AUG' 'SEP' 'OCT' 'NOV' 'DEC'
41     ;
42
43  RUN;
```

Program 17 Notes

Line		
	1	The DATA statement starts the DATA step and names UTLTEMP as the output data set. UTLTEMP is a temporary SAS data set because no permanent SAS library prefix is used. UTLTEMP will be automatically deleted when the SAS session ends. (See Chapter 2 in this book for more on temporary and permanent SAS data sets.)
	2	The KEEP statement lists the variables to keep in the UTLTEMP data set. All other variables are dropped. As a result, UTLTEMP will have three variables: MONTH, TYPE, and AMOUNT. (See End Notes for a complete listing of the UTLTEMP data set.)
	3	The SET statement reads the EXPENSES.UTILITY data set one observation at a time.
	5-11	These lines make up two groups of three statements each. Each group creates an observation for either a gas or an electric expenditure.
	5	This assignment statement sets the value of the variable TYPE to the text string "E". TYPE serves as a tag on the source of the expense. "TYPE" is an arbitrary variable name.
	6	AMOUNT is assigned the value of the variable ELEC–the electricity expense for the current month.
	7	The OUTPUT statement writes an observation to SAS data set UTLTEMP. The KEEP statement (line 2) specifies that the only variables written will be MONTH, TYPE, and AMOUNT. At this point, the value of MONTH is the value just read from EXPENSES.UTILITY. The value of TYPE was assigned in line 5 above as the literal text "E", and the value of AMOUNT was assigned in line 6 as the electricity cost for the current month.

Line 7 completes the group of statements that pulls out the electric cost information and sends it, as one observation, to UTLTEMP. |
| | 9 | This line begins the group of statements (lines 9, 10, and 11) that pulls the information for gas cost. Here, TYPE is assigned the text string "G". |
| | 10 | This assignment statement sets the value of AMOUNT to the current gas cost as held in the variable GAS from EXPENSES.UTILITY. |

Program 17 Notes

Line 11 At this point TYPE is set to "G", AMOUNT is set to the current gas cost, and MONTH to the current month as read from EXPENSES.UTILITY. The OUTPUT statement writes these values as the next observation to UTLTEMP. This is the second observation written on this loop through the DATA step.

Control now jumps back to the top of the DATA step and the start of the next loop. The SET statement is executed again to read the next observation from EXPENSES.UTILITY. For each observation read from EXPENSES.UTILITY, two observations are written to UTLTEMP–one by the OUTPUT statement on line 7, and one by the OUTPUT statement on line 11.

12 The RUN statement ends and executes the DATA step.

14 The DEVICE= option in the GOPTIONS statement names the graphics output device: "HP7440", a Hewlett-Packard 7440 plotter. See Chapter 16 for more on graphics devices.

The ROTATE= option specifies page orientation: LANDSCAPE or PORTRAIT.

16 The bars in the chart are divided into two parts: one indicating the electric cost, and one the gas cost. These two parts will be distinguished by two different fill patterns. The PATTERN1 statement specifies the first pattern to be used. The VALUE= option sets a SOLID fill pattern. The color will be black as set in the COLOR=BLACK option.

17 The PATTERN2 statement defines the second fill pattern as VALUE=EMPTY. The empty bar will be outlined in black as specified in the COLOR=BLACK option.

19-20 The first AXIS statement begins here and ends with the semicolon on line 20. You can define up to 99 sets of AXIS definitions (AXIS1-AXIS99). This set is AXIS1. AXIS is a global statement, meaning it is independent of any one procedure step. In Program 17, the AXIS1 settings affect the chart's vertical (response) axis with the RAXIS= option (line 36). The AXIS statement LABEL= option defines the axis label text. Here, LABEL= specifications call for an axis label text size of 5% of the total graphics area (HEIGHT= 5 PCT), the SWISS typeface (FONT=SWISS), and "Total Cost" as the label text. VALUE= specifications are applied to tick mark labels. These are the dollar amounts on the vertical axis. The VALUE= text size is 2.5% of the graphics area, and the font is SWISS. Note the use of parentheses to group the LABEL= and VALUE= settings. (See End Notes for more on the AXIS statement.)

21-22 This AXIS statement establishes a second set of axis definitions. The AXIS2 settings affect the horizontal (midpoints) axis with the MAXIS= option (line 37). Like the AXIS1 statement above, settings are specified for both the axis label and the tick mark labels.

Program 17 Notes

Line 24-26 The LEGEND statement settings apply to the legend that distinguishes the gas expenses from the electric expenses. Like the AXIS statement, LEGEND is global. It could be used by any number of following procedures. You can define up to 99 sets of definitions (LEGEND1-LEGEND99). The LABEL= option applies to the overall legend label text. The syntax is the same as the LABEL= option in the AXIS statement. Here, the legend will be labeled "Source:" with a text size of 5% of the graphics area, and the SWISS typeface will be used. The VALUE= option applies to the text of legend items. (See End Notes for more on the LEGEND statement.)

28 The PROC statement begins the GCHART procedure and names the UTLTEMP data set as input. UTLTEMP is the temporary SAS data set created in the previous DATA step.

30 The TITLE statement applies to the PROC GCHART output. The quoted text will appear at the top of each page of output. Text size is 8% of the graphics area. The typeface is SWISS.

31 The FORMAT statement applies the SAS format DOLLAR5. to the display of AMOUNT variable values on the vertical axis. DOLLAR5. will show a maximum of five characters including dollar signs and commas. No decimal places are specified. (See End Notes in Chapter 15 for more on the DOLLAR format.)

33 The VBAR (vertical bar) statement begins on this line and ends with the semicolon on line 41. MONTH is the charting variable, meaning there will be a bar for each unique value of MONTH. The "/" separates the charting variable from the options on the following lines.

34 The SUBGROUP=*variable* option can be interpreted as: *Show a bar segment for each value of variable-name.* The variable TYPE was assigned either a "G" or an "E" in the preceding DATA step, so SUBGROUP=TYPE means: *Show one bar segment for "G" values and one for "E" values.* This creates a stacked, two-part, bar for each month.

Closer Look

PATTERN, AXIS, and LEGEND are global statements. Once assigned, these settings apply to all subsequent steps until explicitly changed. For example, the statement **PATTERN1 VALUE=X3 COLOR=RED;** sets the fill pattern to the X3 crosshatch and the fill color to red. If you later use the statement **PATTERN1 VALUE=SOLID;** the fill pattern changes to solid, but the fill color will still be red. It will remain red until you change it. Note that if you specify color as black and are previewing your graphic on a display with a black background, you will not see the graphic. To avoid this problem, use the GOPTIONS statement TARGETDEVICE= option. See "Previewing Graphics Output on a Different Output Device" in Chapter 12, "The GOPTIONS Statement" in *SAS/GRAPH Software: Reference, Version 6, First Edition, Volume 1*.

You can reset global settings to their default definitions with the RESET= graphics option in a GOPTIONS statement. To reset all PATTERN definitions, use **GOPTIONS RESET=PATTERN.** See "RESET" in Chapter 5, "Graphics Options and Device Parameters Dictionary," *SAS/GRAPH Software: Reference, Volume 1*.

Program 17 Notes

Line **35** The SUMVAR=*variable* option means: *Make the height of each bar represent the total of variable-name.* SUMVAR=AMOUNT causes each bar to show the total of AMOUNT, that is, the total of the electricity and gas expense. (See Chapter 16 for more on the effect of the SUMVAR= option.)

36 The RAXIS= option applies the AXIS1 settings (lines 19 and 20) to the response axis. For vertical bar charts, the response axis is the vertical axis.

37 The MAXIS= option applies the AXIS2 settings (lines 21 and 22) to the midpoints axis. For vertical bar charts, the midpoints axis is the horizontal axis.

38 The LEGEND= option applies the LEGEND1 settings (lines 24-26) to the chart legend.

39 The MIDPOINTS= option forces PROC GCHART to arrange the bars in the order listed. There will be one bar for each month, and you want them listed in calendar order. Without the MIDPOINTS= option, the months would be listed in alphabetical order: "APR", "AUG", "DEC", etc.

41 The semicolon ends the VBAR statement that begins on line 33.

43 The RUN statement executes the PROC GCHART procedure.

Closer Look

Program 16, in the previous chapter, uses the DISCRETE option to force PROC GCHART to show one bar for each unique value of the charting variable. **The DISCRETE option is only meaningful when the charting variable is numeric.** Here, the charting variable MONTH is character, so PROC GCHART will always attempt to show one bar for each unique value, no matter how many there are. The DISCRETE option is not needed in Program 17.

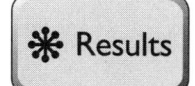

✱ Results | ## Utility Expense Chart

The utility expense chart appears below. There is one bar for each value of MONTH, and each bar is broken into TYPE segments "G" and "E" by the SUBGROUP=TYPE option. These segments are labeled in the legend as defined in the LEGEND statement (lines 19-20) and applied with the LEGEND= option (line 38). The SOLID and EMPTY patterns (PATTERN1 and PATTERN2) distinguish the segments. Vertical axis values are displayed in the DOLLAR5. format as called for by the FORMAT statement on line 31. The vertical axis label and tick mark text are defined in the AXIS1 statement (lines 19-20) and applied by the RAXIS= option (line 36). "Total Cost" is substituted for the variable name "AMOUNT" as the vertical axis label. The horizontal axis is likewise controlled by the AXIS2 statement (lines 21-22) and the MAXIS= option (line 37).

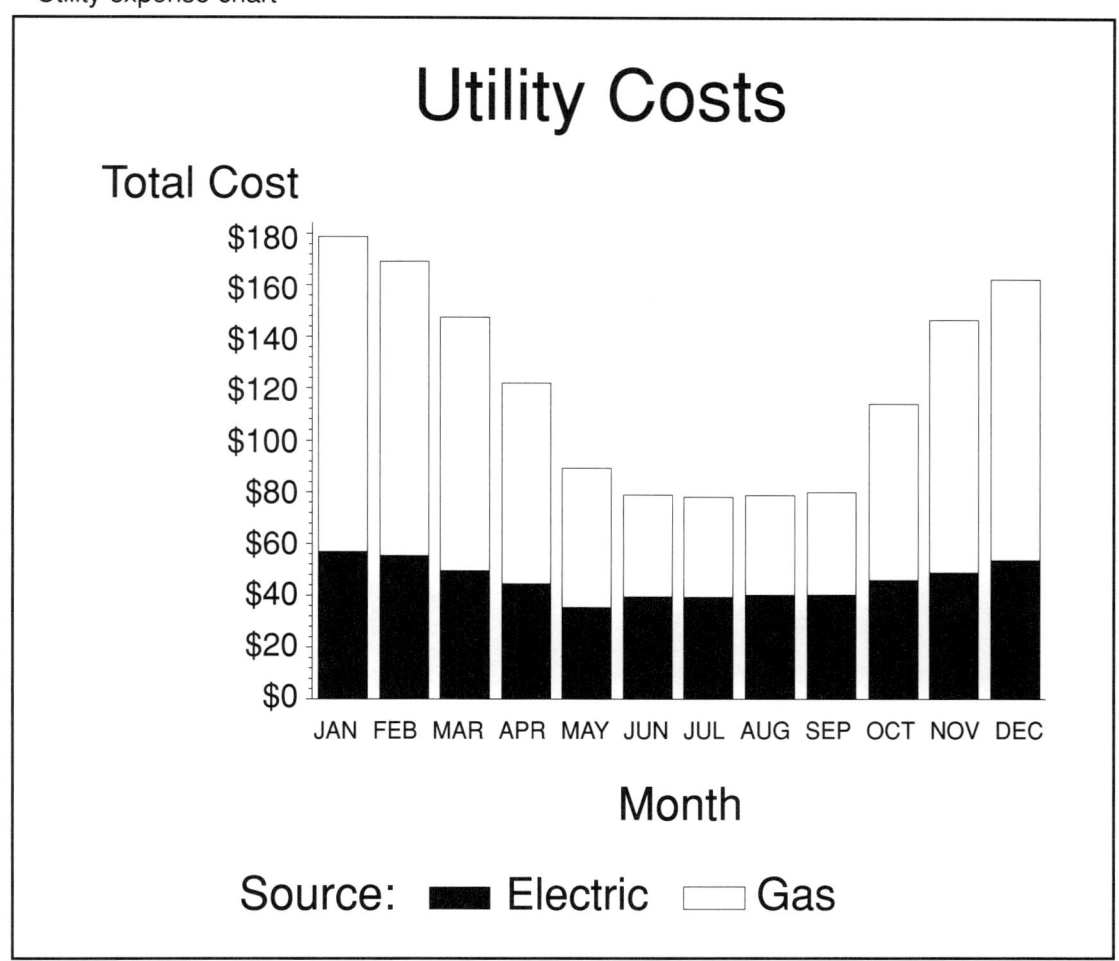

Utility expense chart

Chapter 17 - Creating Stacked Bar Charts

Data Organization, Chart Text

Data Organization. The DATA step in Program 17 restructures EXPENSES.UTILITY into the temporary data set UTLTEMP. This restructuring is required to conform to the data layout expected by PROC GCHART. When you use the SUBGROUP= and SUMVAR= options, the charting data set must have at least three variables:

- the charting variable named after VBAR (MONTH in Program 17)
- a SUMVAR= variable (AMOUNT in Program 17)
- a SUBGROUP= variable (TYPE in Program 17).

There will be one bar for each value of the charting variable. Each bar will be made up of a segment for each unique value of the SUBGROUP= variable. Each of the these segments will represent the value of the SUMVAR= variable. The DATA step in Program 17 creates the required variables in UTLTEMP. The complete UTLTEMP data set appears below.

Temporary data set UTLTEMP

OBS	MONTH	TYPE	AMOUNT
1	JAN	E	56.76
2	JAN	G	122.34
3	FEB	E	55.23
4	FEB	G	114.08
5	MAR	E	49.61
6	MAR	G	98.25
7	APR	E	43.98
8	APR	G	77.67
9	MAY	E	40.77
10	MAY	G	48.01
11	JUN	E	39.07
12	JUN	G	39.80
13	JUL	E	39.82
14	JUL	G	37.91
15	AUG	E	39.74
16	AUG	G	38.47
17	SEP	E	40.27
18	SEP	G	41.35
19	OCT	E	45.34
20	OCT	G	68.50
21	NOV	E	48.99
22	NOV	G	97.72
23	DEC	E	53.70
24	DEC	G	108.83

▸ **The AXIS statement.** With the AXIS statement you can control the appearance of axis and tick mark labels. The LABEL= option settings apply to the axis label. VALUE= option settings apply to the tick mark labels. Also available are several other options and settings not used in Program 17.

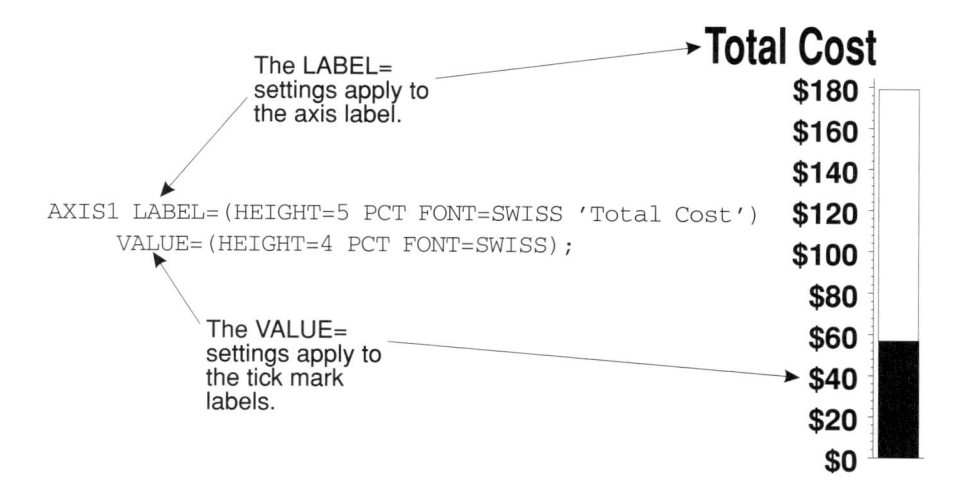

The LABEL= settings apply to the axis label.

```
AXIS1 LABEL=(HEIGHT=5 PCT FONT=SWISS 'Total Cost')
      VALUE=(HEIGHT=4 PCT FONT=SWISS);
```

The VALUE= settings apply to the tick mark labels.

Total Cost
$180
$160
$140
$120
$100
$80
$60
$40
$20
$0

▸ **The LEGEND statement.** The LEGEND statement gives you considerable control over the appearance of chart legends. In Program 17, LEGEND settings are used to control the look and content of the text that distinguishes the gas costs from the electric costs. You can use other LEGEND settings to control color, position, and arrangement of legend text. In addition, you can add drop shadows and frames to the legend. The VBAR statement LEGEND= option is only meaningful when you use the SUBGROUP= option to create stacked bar charts.

To control the legend label, use the LEGEND statement LABEL= option. Use the VALUE= option to control the text of legend items. (See graphic below.) When using the VALUE= option, the first specification applies to the first (far left) item. In Program 17 the first item is the electricity portion of the stacked bar, where the value of TYPE is "E". The second set of specifications applies to the second item. Legend items are arrayed alphabetically. The electricity cost is first (far left) because "E" comes alphabetically before "G".

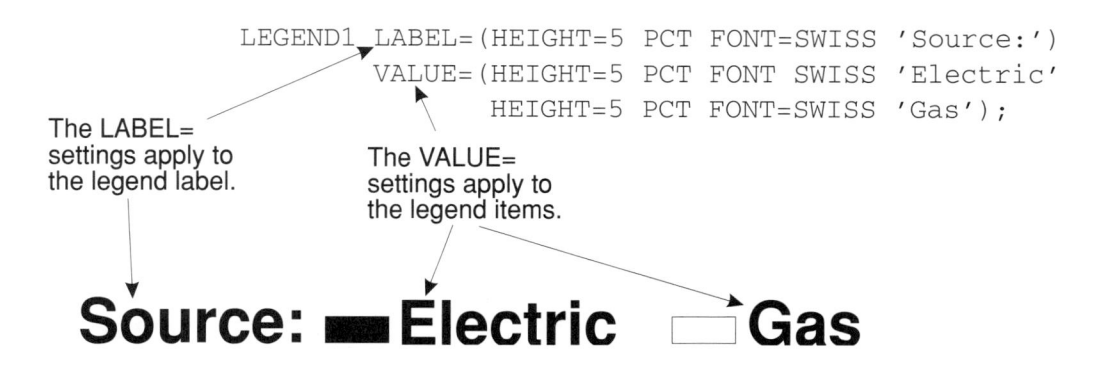

```
LEGEND1 LABEL=(HEIGHT=5 PCT FONT=SWISS 'Source:')
        VALUE=(HEIGHT=5 PCT FONT SWISS 'Electric'
               HEIGHT=5 PCT FONT=SWISS 'Gas');
```

The LABEL= settings apply to the legend label.

The VALUE= settings apply to the legend items.

Source: ■ Electric ☐ Gas

Chapter 17 - Creating Stacked Bar Charts **191**

📝 **Default text settings.** In Program 17, the AXIS, LEGEND, FORMAT, and LABEL statements, plus the MAXIS=, RAXIS=, LEGEND=, FONT=, and HEIGHT= options control the appearance of text on the chart. You can simplify your charting program by accepting the defaults for axis and title text. The GCHART procedure step below is similar to the step in Program 17, except that the text control statements and options are removed:

```
PATTERN1 VALUE=SOLID COLOR=BLACK;
PATTERN2 VALUE=EMPTY COLOR=BLACK;

PROC GCHART DATA=UTLTEMP;

   TITLE 'Utility Costs';

   VBAR MONTH /
      SUBGROUP=TYPE
      SUMVAR=AMOUNT
      MIDPOINTS = 'JAN' 'FEB' 'MAR' 'APR' 'MAY' 'JUN'
                  'JUL' 'AUG' 'SEP' 'OCT' 'NOV' 'DEC'
   ;
RUN;
```

The resulting chart appears below. Notice the default text settings for the title, axis, and legend. Also note that without an alternative legend setting, the values of the variable TYPE ("E" and "G") are used to label legend items. The name of the SUMVAR= variable (AMOUNT) and the statistic the bars represent (SUM) are used to label the vertical axis.

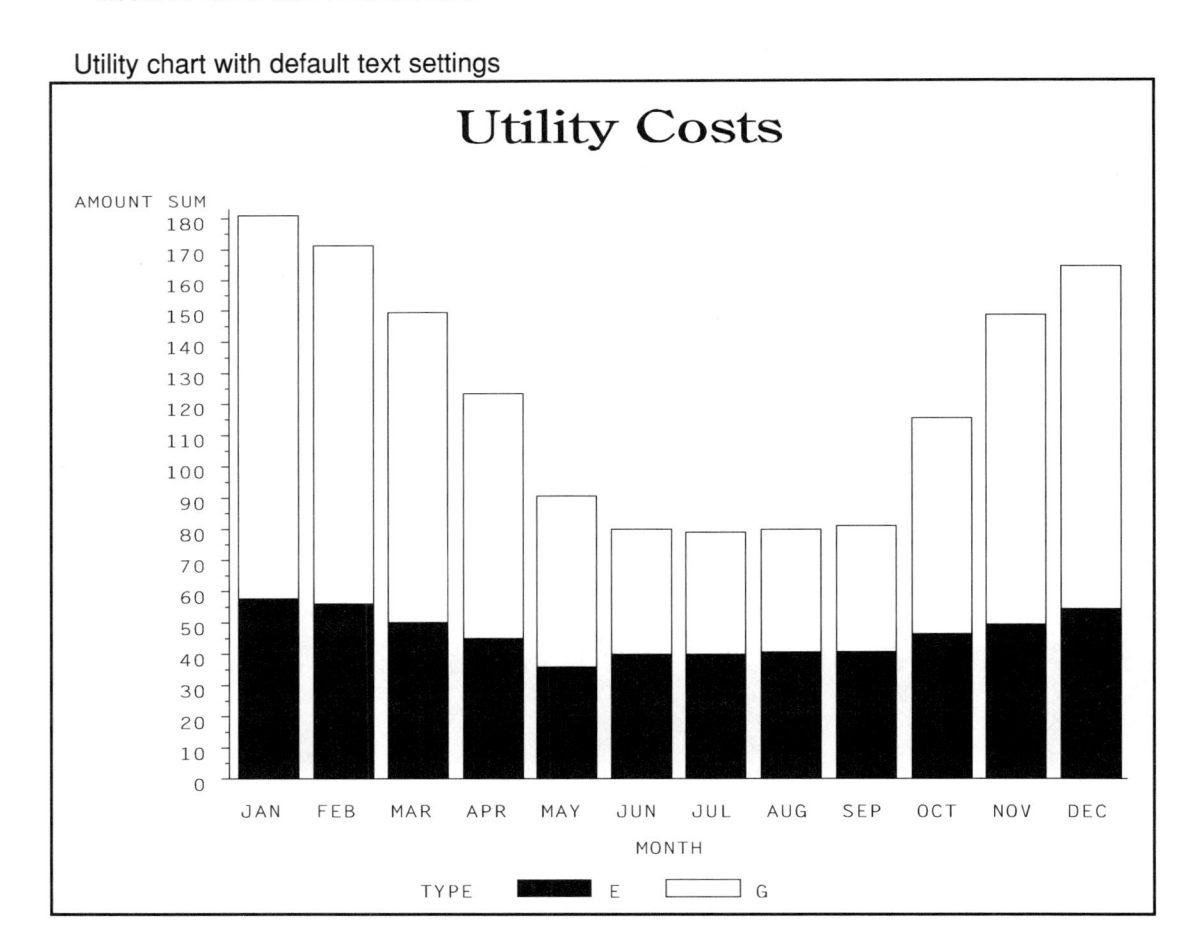

Utility chart with default text settings

192 Part 3 - Presenting Your Data

Creating Stacked Bar Charts

1. Make sure you know the name and have access to the SAS data set with which you want to work. This may require a LIBNAME statement. (See Chapter 2 in this book for a LIBNAME example.)

2. The data set you want to chart from must be organized properly for use with the SUBGROUP= option in PROC GCHART. You must have a separate SUBGROUP variable.

3. Make sure you know the name of the graphics device driver where your chart will be printed or displayed. Name the output device with the GOPTIONS DEVICE= option. Use PROC GDEVICE to see a list device drivers.

4. Use PROC GCHART. Name the variable you want to chart in the VBAR statement. You get a bar for each value of the charting variable.

5. Use the SUBGROUP= option in the VBAR statement to name the variable that distinguishes the bar segments. Precede VBAR options with a "/".

6. Use the AXIS, PATTERN, and LEGEND statements to control the appearance of the chart.

Charting Features and the OUTPUT Statement

AXIS statement
"Modifying the Axes" in Chapter 6, "Producing Vertical Bar Charts," *SAS/GRAPH Software: Usage, Version 6, First Edition;*
Chapter 9, "The AXIS Statement," *SAS/GRAPH Software: Reference, Version 6, First Edition, Volume 1*

LEGEND statement
"Modifying the Legend" in Chapter 7, "Grouping and Subgrouping," *SAS/GRAPH Software: Usage;*
Chapter 13, "The LEGEND Statement," *SAS/GRAPH Software: Reference, Volume 1*

OUTPUT statement
"Creating Multiple Observations from a Single Observation" in Chapter 9, "Reshaping Data," *SAS Language and Procedures: Usage 2, Version 6, First Edition;*
"OUTPUT" in Chapter 9, "SAS Language Statements," *SAS Language: Reference, Version 6, First Edition*

PATTERN statement
"Changing the Patterns of the Bars" in Chapter 5, "Producing Horizontal Bar Charts," *SAS/GRAPH Software: Usage;*
Chapter 15, "The PATTERN Statement," *SAS/GRAPH Software: Reference, Version 6, First Edition, Volume 2*

PROC GCHART general reference
Chapter 23, "The GCHART Procedure," *SAS/GRAPH Software: Reference, Volume 2*

SAS graphics options (GOPTIONS)
Chapter 5, "Graphics Options and Device Parameters Dictionary," *SAS/GRAPH Software: Reference, Volume 1*

VBAR statement and options
Chapter 6, "Producing Vertical Bar Charts," *SAS/GRAPH Software: Usage;*
"VBAR Statement" in Chapter 23, "The GCHART Procedure," *SAS/GRAPH Software: Reference, Volume 2*

Chapter 18
Creating Grouped Bar Charts

Learn How to...

- Create a chart with grouped bars
- Select a graphics output device
- Select fill patterns and text fonts
- Organize data for charting

Using These SAS System Features...

- DATA step
- GCHART procedure
- GROUP= option
- GOPTIONS statement
- PATTERN statement
- AXIS statement

? Problem — Create a Quarterly Profits Chart

The Accounting Department tallies the profits of three corporate divisions for each quarter of the year. The information is stored in SAS data set ACCOUNT.QTRPROF, which appears below. The variable QTR holds the quarter of the year. DIRECT is the profit for the Direct Sales division, RETAIL the profit for the Retail division, and MANUFAC the profit for the Manufacturing division. Profit values are in thousands of dollars. You need to produce a bar chart showing profits for each quarter with a separate bar for each division. Each quarter will have three bars: one for Direct Sales, one for Retail, and one for Manufacturing.

SAS data set ACCOUNT.QTRPROF

```
               Quarterly Division Profits

     OBS     QTR     DIRECT      RETAIL      MANUFAC

      1       1        256          98         1034
      2       2        101         -40          452
      3       3        124          23          601
      4       4        212          76          929
```

Solution: GCHART Procedure and VBAR Statement

The SAS/GRAPH GCHART procedure allows you to create the profits chart as a grouped bar chart. Before you can use PROC GCHART, your data must be organized properly. Program 18 generates the required chart with two SAS steps. First, a DATA step reorganizes the data into the temporary SAS data set named PROFTEMP. The temporary data set is structured so that each observation represents the quarterly profit for a single division. Next, a PROC GCHART step with the GROUP= option creates the chart. The GROUP= option generates a "minichart" for each value of the GROUP= variable. Here, that variable is QTR. There are four minicharts–one for each quarter.

Program 18

```
1   DATA PROFTEMP;
2
3      KEEP QTR DIVISION PROFIT;
4      SET ACCOUNT.QTRPROF;
5
6      DIVISION = 'D';
7      PROFIT= DIRECT;
8      OUTPUT;
9
10     DIVISION = 'R';
11     PROFIT= RETAIL;
12     OUTPUT;
13
14     DIVISION = 'M';
15     PROFIT= MANUFAC;
16     OUTPUT;
17
18  RUN;
19
```

(continued)

Program 18 (continued)

```
20  GOPTIONS DEVICE=HP7440 ROTATE=LANDSCAPE;
21
22  AXIS1 LABEL=(HEIGHT=4 PCT FONT=SWISS 'Net Profit')
23        VALUE=(HEIGHT=3 PCT FONT=SWISS);
24  AXIS2 LABEL=(HEIGHT=4 PCT FONT=SWISS 'Quarter')
25        VALUE=(HEIGHT=3 PCT FONT=SWISS);
26  AXIS3 LABEL=(HEIGHT=4 PCT FONT=SWISS 'Division')
27        VALUE=(HEIGHT=3 PCT FONT=SWISS);
28
29  PATTERN1 VALUE=SOLID COLOR=BLACK;
30
31  PROC GCHART DATA=PROFTEMP;
32
33     TITLE HEIGHT=6 PCT FONT=SWISS 'Quarterly Profits';
34     TITLE2 HEIGHT=4 PCT FONT=SWISS
35            '(in thousands of dollars)';
36     FOOTNOTE HEIGHT=3 PCT FONT=ZAPF
37       'D=Direct Marketing   M=Manufacturing   R=Retail';
38
39     FORMAT PROFIT DOLLAR5.;
40
41     VBAR DIVISION /
42        GROUP=QTR
43        SUMVAR=PROFIT
44        RAXIS=AXIS1
45        GAXIS=AXIS2
46        MAXIS=AXIS3
47     ;
48
49  RUN;
```

Closer Look

Program 18 can be greatly simplified if you leave out the AXIS statements, the VBAR axis options, the PATTERN statement, and font and typeface specifications for titles and footnotes. The resulting chart will be built using defaults for chart text. See End Notes in Chapter 17 for an example.

Program 18 Notes

Line   **1**   The DATA statement starts the DATA step and names PROFTEMP as the output data set. Because it has a one-level name with no SAS library prefix, PROFTEMP is a temporary SAS data set. PROFTEMP will be automatically deleted when the SAS session ends. (See Chapter 2 in this book for more on temporary and permanent SAS data sets.)

3   The KEEP statement lists the variables to keep in the output data set, PROFTEMP. PROFTEMP will contain only three variables: QTR, DIVISION, and PROFIT. The variables DIRECT, RETAIL, and MANUFAC are dropped.

4   The SET statement reads the ACCOUNT.QTRPROF data set one observation at a time.

6-16   These lines make up three groups of three statements each. For each observation read from ACCOUNT.QTRPROF, these statements generate three output observations, one for each of the three corporate divisions.

6   This assignment statement sets the value of DIVISION to the character "D", indicating the Direct Sales division.

7   The PROFIT variable is used to hold the net profit for each division. Here, PROFIT is set to the value of DIRECT as read from ACCOUNT.QTRPROF.

8   The OUTPUT statement writes an observation to the output SAS data set, PROFTEMP. The KEEP statement (line 3) specifies that the only variables written will be QTR, DIVISION, and PROFIT. The current value of QTR is the value read from ACCOUNT.QTRPROF in line 4. The value of DIVISION was assigned as "D" in line 6 above, and the value of PROFIT was assigned in line 7 as the current value of the variable DIRECT.

Line 8 completes the group of statements (lines 6, 7, and 8) that gathers information for the Direct Sales division for the quarter and sends it, as one observation, to PROFTEMP.

10   This line begins the group of statements (lines 10, 11, and 12) that gathers the information for the Retail division. Here, DIVISION is assigned the character "R".

11   This assignment statement sets the value of PROFIT to the profit for the Retail division held in RETAIL.

Program 18 Notes

Line **12** At this point, DIVISION is set to "R", PROFIT to the quarterly profit for Retail, and QTR is still the current quarter as read by the SET statement in line 4. This OUTPUT statement writes these values as the next observation to PROFTEMP.

14 The variable DIVISION is set to "M", indicating the Manufacturing division.

15 PROFIT is set to the current quarterly profit for Manufacturing held in MANUFAC.

16 This OUTPUT statement writes the Manufacturing information to PROFTEMP. DIVISION and PROFIT values have been set in lines 14 and 15, and the value of QTR remains the value read from ACCOUNT.QTRPROF in line 4. This is the third observation written to PROFTEMP on this loop through the DATA step.

Control now jumps to the top of the DATA step. The SET statement (line 4) is executed again to read in the next observation.

18 The RUN statement ends and executes the DATA step.

20 The DEVICE= option in the GOPTIONS statement names the graphics output device: "HP7440", a Hewlett-Packard 7440 plotter. See Chapter 16 for more on graphics devices.

The ROTATE= option specifies page orientation: LANDSCAPE or PORTRAIT.

22-23 This AXIS statement defines the AXIS1 settings. These settings are applied to the horizontal (response) axis by the RAXIS= option on line 44. The axis label is assigned the text "Net Profit", a size of 4% of the graphics area (HEIGHT= 4 PCT), and the SWISS typeface (FONT=SWISS). Tick mark labels are 3% of the graphics area and are displayed in the SWISS typeface. (See End Notes in Chapter 17 for more on the AXIS statement.)

24-25 This AXIS statement defines the AXIS2 settings. The AXIS2 definitions are applied to the grouping variable text by the GAXIS= option on line 45. The label is assigned the text "Quarter", size is 4% of the graphics area, and the typeface is SWISS. Values are displayed at 3% of the graphics area, in the SWISS typeface. (See End Notes for more on how the AXIS settings affect the display of the grouping variable.)

26-27 The AXIS3 settings defined here are applied to the horizontal (midpoints) axis by the MAXIS= option on line 46. The horizontal axis shows values of the charting variable, DIVISION. The horizontal axis is labeled "Division" at 4% of the graphics area, in the SWISS typeface. The DIVISION variable values are displayed at 3% of the graphics area in the SWISS typeface.

Program 18 Notes

Line

29 The PATTERN1 statement defines the bar fill pattern for your chart. The VALUE=SOLID option sets a solid fill. Fill color is black.

31 The PROC statement begins the GCHART procedure and names the PROFTEMP data set as input. PROFTEMP is the temporary SAS data set created in the previous DATA step.

33 The TITLE statement applies to the PROC GCHART output. The title text size is set to 6% of the graphics area, the typeface to SWISS, and the text itself to "Quarterly Profits". Note that TITLE is equivalent to TITLE1. Either defines settings for the first title line.

34-35 The TITLE2 statement defines the second line of title text. The second title line is smaller than the first, 4% of the graphics area versus 6%. The typeface is SWISS. The TITLE statement, like all SAS statements, can span multiple lines.

36-37 The FOOTNOTE statement defines text that will appear at the bottom of each page of output. The ZAPF typeface is specified (FONT=ZAPF), and size is set to 3% of the graphics area.

39 The FORMAT statement applies the SAS format DOLLAR5. to the display of PROFIT variable values. DOLLAR5. will show a maximum of five characters including dollar signs and commas, but no decimal places. (See Chapter 15 in this book for more on the DOLLAR*n*. format.)

Closer Look

The PATTERN statement controls the bar fill pattern and color. **When the COLOR= option is not used the fill color will be selected from a default** *color list.* For the HP7440 device, the first color in the default list is black, so `PATTERN1 VALUE=SOLID;` produces the same results as `PATTERN1 VALUE=SOLID COLOR=BLACK;`.

A default color list is defined for each graphics output device. For most hardcopy devices the first color selected will be black; for most display devices it will be white. If you specify color as black and are previewing your graphic on a display with a black background, you will not see the graphic. To avoid this problem use the GOPTIONS statement TARGETDEVICE= option. See "Previewing Graphics Output on a Different Output Device" in Chapter 12, "The GOPTIONS Statement," and Chapter 7, "SAS/GRAPH Colors" in *SAS/GRAPH Software: Reference, Version 6, First Edition, Volume 1.*

Program 18 Notes

Line **41** The VBAR (vertical bar) statement begins on this line and ends on line 47. DIVISION is the charting variable, meaning there will be a bar for each unique value of DIVISION. VBAR options follow the "/".

42 The GROUP=*variable* option can be interpreted as: *Produce a minichart for each value of variable-name.* The variable QTR has four unique values ("1", "2", "3", and "4"), so there will be four minicharts or groups. Each group will have three bars, one for each value of the charting variable DIVISION ("D", "R", and "M").

43 The SUMVAR=*variable* option means: *Make the height of each bar represent the sum of variable-name.* SUMVAR=PROFIT causes each bar to show the PROFIT for each division. See Chapter 16 in this book for more on the effect of the SUMVAR= option.

44 The RAXIS= option applies the AXIS1 definitions to the response axis. For vertical bars charts the response axis is the vertical axis.

45 The GAXIS= option applies the AXIS2 definitions to the grouping variable text. The grouping variable is QTR. The grouping variable label and values will be displayed with the definitions assigned with the AXIS2 statement options (line 24).

46 The MAXIS= option applies the AXIS3 definitions to the midpoints axis. For vertical bars charts the midpoints axis is the horizontal axis.

47 The semicolon ends the VBAR statement started on line 41.

49 The RUN statement executes the GCHART procedure.

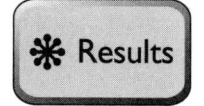

Results — Quarterly Profits Chart

The quarterly profits chart appears below. A minichart is generated for each value of the GROUP= variable, QTR. The height of the bars represents the SUMVAR= variable, PROFIT. All bars are solid filled, as called for by the PATTERN1 statement (line 29). Text assigned in the LABEL= and VALUE= options (in the AXIS statements) is used instead of the variable names "QTR", "DIVISION", and "PROFIT". As this example demonstrates, you can use footnotes and titles to add explanations and customized legends to your chart.

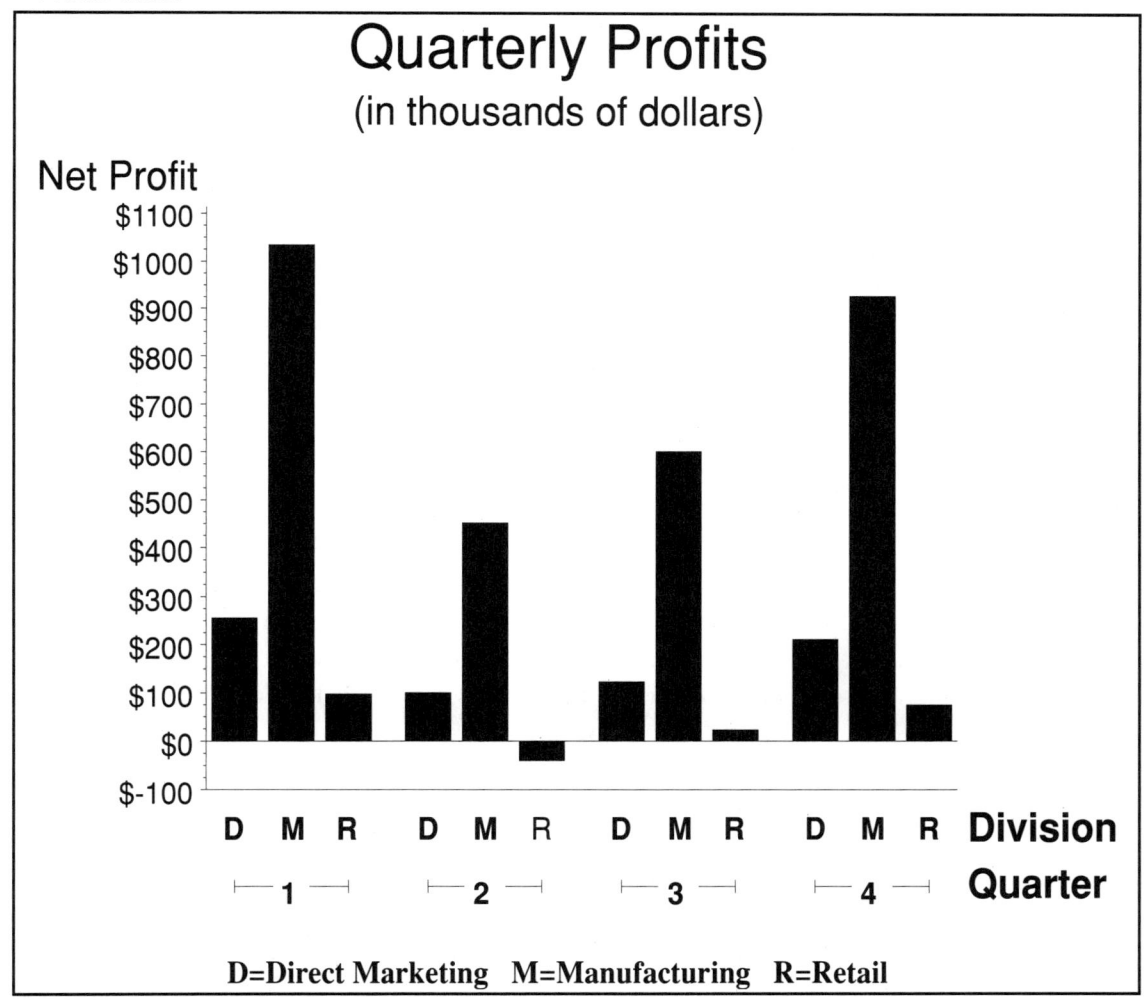

Profits chart

Quarterly Profits
(in thousands of dollars)

D=Direct Marketing   M=Manufacturing   R=Retail

Chapter 18 - Creating Grouped Bar Charts

Controlling Chart Appearance

📝 **Group AXIS settings.** You can assign a set of axis definitions to the grouping variable with the GAXIS= option (line 45 in Program 18). Apply the LABEL= and VALUE= option settings as shown below:

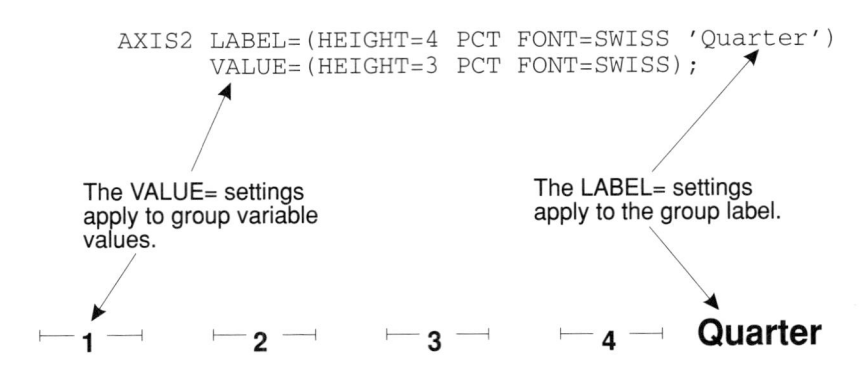

```
AXIS2 LABEL=(HEIGHT=4 PCT FONT=SWISS 'Quarter')
      VALUE=(HEIGHT=3 PCT FONT=SWISS);
```

The VALUE= settings apply to group variable values.

The LABEL= settings apply to the group label.

⊢ 1 ⊣   ⊢ 2 ⊣   ⊢ 3 ⊣   ⊢ 4 ⊣   **Quarter**

📝 **The NOTE statement.** In addition to TITLE and FOOTNOTE text, you can also use the NOTE statement to place text anywhere in the procedure output area of your chart. The NOTE statement syntax is similar to that for the TITLE statement. With NOTE you can place annotation text on your chart, as well as draw lines and boxes. Unlike TITLE and FOOTNOTE statements, which are globals, NOTE statements are local. This means they apply only to the procedure in which they appear.

NOTE, TITLE, FOOTNOTE, and other SAS/GRAPH statements give you considerable latitude to customize your graphics, but be prepared to spend some time experimenting. Using these statements to customize SAS/GRAPH output is an iterative process of writing statements, running the procedure, checking the output, and changing statements if necessary. This can be time consuming. The advantage of this noninteractive method comes when you are repeatedly generating many charts. Once you set up notes, titles, and other graphic elements, SAS/GRAPH software can handle large amounts of data with no further intervention. See Chapter 14, "The NOTE Statement" in *SAS/GRAPH Software: Reference, Volume 1*.

GROUP= and SUBGROUP=. Chapter 17 in this book demonstrates the SUBGROUP= option. SUBGROUP= creates stacked bar charts. PROC GCHART makes it easy to switch between SUBGROUP= and GROUP= chart styles, so you can experiment to find the chart type that best expresses your data. In the example below, Program 18 is modified to generate a stacked bar chart instead of a grouped bar chart. Three patterns are defined to distinguish between the three divisions that make up each bar. A LEGEND statement and LEGEND= option in the VBAR statement create a customized legend. Note that, in this case, a stacked bar chart does not accurately represent net profit because the negative result for the Retail division in the second quarter is not subtracted from the bar height. Instead, it is displayed as a negative bar. The resulting chart appears on the next page.

```
AXIS1 LABEL=(HEIGHT=4 PCT FONT=SWISS 'Net Profit')
      VALUE=(HEIGHT=3 PCT FONT=SWISS);
AXIS2 LABEL=(HEIGHT=4 PCT FONT=SWISS 'Quarter')
      VALUE=(HEIGHT=3 PCT FONT=SWISS);

LEGEND1 LABEL=(HEIGHT=4 PCT FONT=SWISS 'Division: ')
        VALUE=(HEIGHT=3 PCT FONT=SWISS 'Direct Market'
               HEIGHT=3 PCT FONT=SWISS 'Manufacturing'
               HEIGHT=3 PCT FONT=SWISS 'Retail');

PATTERN1 VALUE=SOLID COLOR=BLACK;
PATTERN2 VALUE=X3 COLOR=BLACK;
PATTERN3 VALUE=EMPTY COLOR=BLACK;

PROC GCHART DATA=PROFTEMP;
   TITLE HEIGHT=6 PCT FONT=SWISS 'Quarterly Profits';
   TITLE2 HEIGHT=4 PCT FONT=SWISS
          '(in thousands of dollars)';
   FORMAT PROFIT DOLLAR5.;
   VBAR QTR /
      SUBGROUP=DIVISION
      SUMVAR=PROFIT
      RAXIS=AXIS1
      MAXIS=AXIS2
      LEGEND=LEGEND1
   ;
RUN;
```

A stacked bar chart of division profits

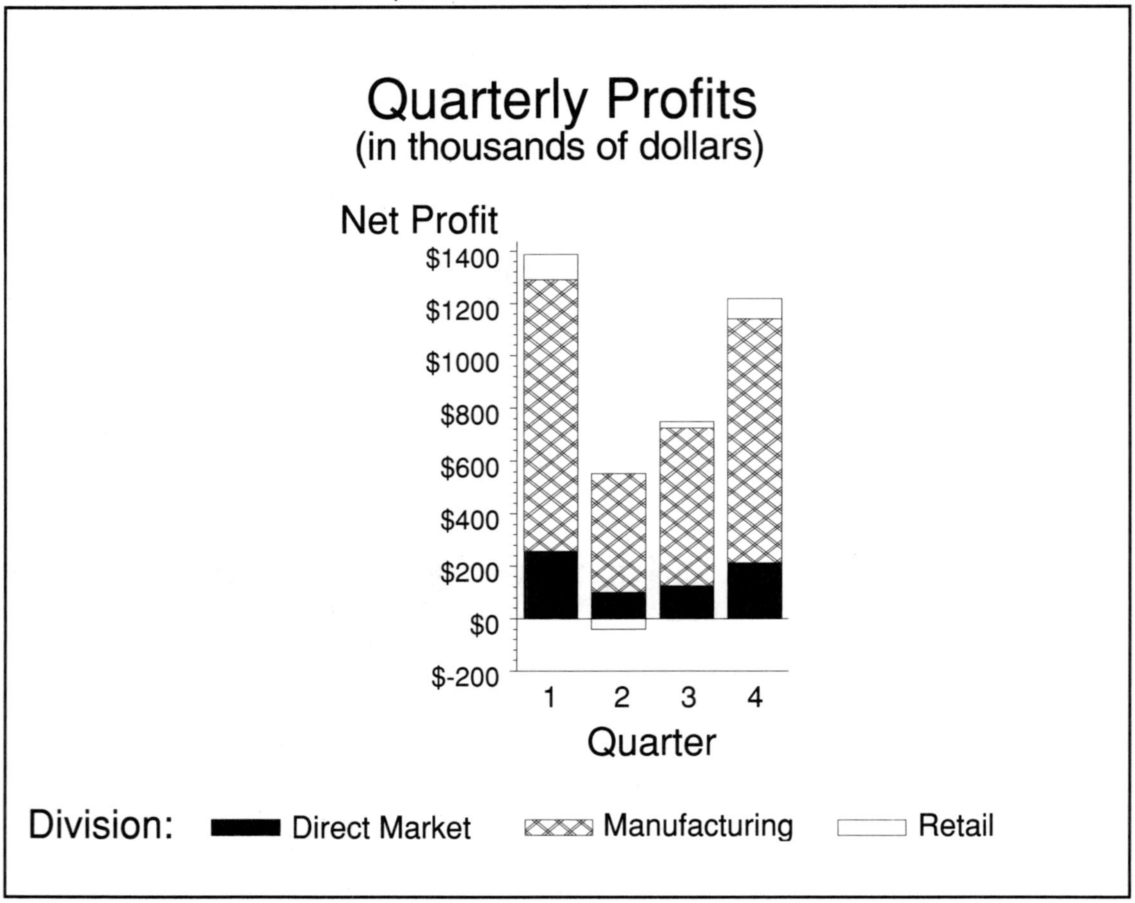
Quarterly Profits
(in thousands of dollars)
Net Profit
$1400
$1200
$1000
$800
$600
$400
$200
$0
$-200
1
2
3
4
Quarter
Division:
Direct Market
Manufacturing
Retail

You can use the SUBGROUP= and GROUP= options together to produce a grouped chart with stacked bars. This can be useful when you need to deal with two category levels or groups within groups. See "Grouping and Subgrouping in the Same Chart" in Chapter 7, "Grouping and Subgrouping," *SAS/GRAPH Software: Usage, Version 6, First Edition.*

Creating Grouped Bar Charts

1. Make sure you know the name and have access to the SAS data set with which you want to work. This may require a LIBNAME statement. (See Chapter 2 in this book for a LIBNAME example.)

2. Make sure you know the name of the graphics device driver where your chart will be printed or displayed. Name the output device with the DEVICE= option in a GOPTIONS statement. Use PROC GDEVICE to list available drivers.

3. The data set you want to chart from must be organized properly for a grouped bar chart. You need a separate grouping variable. The grouping variable is named in the GROUP=*variable* option of the VBAR statement.

4. Use PROC GCHART. Name the variable you want to chart in the VBAR statement. You get a bar for each value of the VBAR variable.

5. Use the GROUP=*variable* option to create a minichart or group for each unique value of *variable*.

6. Use AXIS statements and the VBAR options MAXIS=, RAXIS=, and GAXIS= if you want to customize the appearance of text on the chart.

PROC GCHART, Charting Control Statements

AXIS statement
Chapter 9, "The AXIS Statement," *SAS/GRAPH Software: Reference, Version 6, First Edition, Volume 1*

more GROUP chart examples
Chapter 7, "Grouping and Subgrouping," *SAS/GRAPH Software: Usage, Version 6, First Edition*

PATTERN statement
"Specifying How Patterns Are Assigned" in Chapter 5, "Producing Horizontal Bar Charts," *SAS/GRAPH Software: Usage;*
Chapter 15, "The PATTERN Statement," *SAS/GRAPH Software: Reference, Volume 1*

PROC GCHART general reference
Chapter 23, "The GCHART Procedure," *SAS/GRAPH Software: Reference, Version 6, First Edition, Volume 2*

SAS/GRAPH fonts
Chapter 6, "SAS/GRAPH Fonts," *SAS/GRAPH Software: Reference, Volume 1*

SAS graphics options (GOPTIONS)
Chapter 5, "Graphics Options and Device Parameters Dictionary," *SAS/GRAPH Software: Reference, Volume 1*

VBAR statement and options
Chapter 6, "Producing Vertical Bar Charts," *SAS/GRAPH Software: Usage;*
"VBAR Statement" in Chapter 23, "The GCHART Procedure," *SAS/GRAPH Software: Reference, Volume 2*

Chapter 19
Creating Line Graphs and Plots

Learn How to...

- Create plots and line graphs
- Control the display of points on a graph
- Control the appearance of plot text

Using These SAS System Features...

- GPLOT procedure
- PLOT statement
- GOPTIONS statement
- AXIS statement

? Problem Create a Fish Count Plot

The Department of Natural Resources has a long-term project to research the effects of phosphorous on game fish. The annual average phosphorous concentration in mg/L (milligrams per liter) in Crystal Lake has been recorded for each year from 1978 to 1993. For the same time period, the number of walleye pike captured in a controlled netting has been recorded. These data reside in SAS data set DNRFISH.PHOSTEST, which appears below. The variable YEAR holds the year of the test, PHOS holds the phosphorous concentration in mg/L, and COUNT holds the fish count. To study the relationship between phosphorous concentration and fish count you need to create a plot of fish count versus phosphorous concentration.

SAS data set DNRFISH.PHOSTEST

OBS	YEAR	PHOS	COUNT
1	1978	0.1130	54
2	1979	0.1280	50
3	1980	0.0900	56
4	1981	0.0770	62
5	1982	0.0820	58
6	1983	0.0780	60
7	1984	0.0740	63
8	1985	0.0610	65
9	1986	0.0590	72
10	1987	0.0530	75
11	1988	0.0520	74
12	1989	0.0700	70
13	1990	0.0560	74
14	1991	0.0480	75
15	1992	0.0440	78
16	1993	0.0511	75

Solution: SORT and GPLOT Procedures

Program 19 uses SAS/GRAPH software's GPLOT procedure to create the required plot. Program 19 also makes use of AXIS, SYMBOL, and TITLE statements to control the appearance of text on the plot. Note that DNRFISH.PHOSTEST is in order by year. Before plotting the data, the data set is sorted by phosphorous concentration (PHOS). Sorting ensures that lines connecting points on the plot are drawn correctly. (See End Notes for a further discussion of this problem.)

Program 19

```
1   PROC SORT DATA=DNRFISH.PHOSTEST OUT=PHOSTEMP;
2      BY PHOS;
3   RUN;
4
5   GOPTIONS DEVICE=HP7440 ROTATE=LANDSCAPE;
6
7   AXIS1 LABEL=(HEIGHT=4 PCT FONT=SWISS 'Fish Count')
8         VALUE=(HEIGHT=3 PCT FONT=SWISS);
9
10  AXIS2 LABEL=(HEIGHT=4 PCT FONT=SWISS
11              'P Concentration mg/L')
12        VALUE=(HEIGHT=3 PCT FONT=SWISS);
13
14  SYMBOL1 INTERPOL=JOIN LINE=1 VALUE=NONE;
15
16  PROC GPLOT DATA=PHOSTEMP;
17
18     TITLE HEIGHT=5 PCT FONT=SWISS
19       'Walleye Pike Count v. Phosphorous Concentration';
20     TITLE2 HEIGHT=4 PCT FONT=SWISS
21       'Data from 1978 to 1993';
22
23     PLOT COUNT*PHOS /
24        VAXIS=AXIS1
25        HAXIS=AXIS2
26     ;
27
28  RUN;
```

Program 19 Notes

Line		
	1	The PROC SORT step sorts DNRFISH.PHOSTEST by phosphorous concentration (variable PHOS). The OUT= option sends the sorted observations to the new data set, PHOSTEMP. Because it has a one-level name with no SAS library prefix attached, PHOSTEMP is a temporary data set. PHOSTEMP will be automatically deleted when the SAS session ends. (See Chapter 2 for more on temporary and permanent SAS data sets.)
	2	The BY statement names the sorting variable, PHOS.
	3	The RUN statement ends and executes the SORT procedure.
	5	The DEVICE= option in the GOPTIONS statement names the output device, "HP7440", a Hewlett-Packard 7440 plotter. See Chapter 16 for more on graphics devices.
		The ROTATE= option specifies page orientation, LANDSCAPE.
	7-8	This AXIS statement defines the AXIS1 settings. These settings are applied to the vertical plot axis by the PLOT statement option VAXIS= on line 24. LABEL= settings are used for the axis label. Text for the label is "Fish Count". This label is 4% of the graphics area (HEIGHT=4 PCT), displayed in the SWISS typeface (FONT=SWISS). Vertical axis tick mark labels are controlled by the VALUE= option. These will be sized at 3% of the graphics area, displayed in the SWISS typeface. (See End Notes in Chapter 17 for more on the AXIS statement.)
	10-12	This AXIS statement defines the AXIS2 settings. The AXIS2 definitions are applied to the plot horizontal axis by the HAXIS= option (line 25). The axis label text is "P Concentration mg/L". Size is 4% of the graphics area and the typeface is SWISS. Tick mark values are 3% of the graphics area, in the SWISS typeface.

Closer Look

When working on a plot or other graphic you usually want to check your output on a display before taking the time to send it to a printer. SAS/GRAPH software supports many graphics displays. Displays are treated as separate devices with their own color lists and sizing parameters. Graphics sent to a display device will not look exactly the same when sent to a hardcopy device. To get a closer approximation of what hardcopy output will look like, use the TARGETDEVICE= graphics option, for example:

```
GOPTIONS DEVICE=PMVGA TARGETDEVICE=HP7440;
```

This GOPTIONS statement sets the graphics output device to a VGA display (valid when running under OS/2). The target device is set to the Hewlett-Packard 7440 plotter. After executing this statement, SAS/GRAPH output will go to the VGA display and will appear as it would on the plotter.

Program 19 Notes

Line 14 — This SYMBOL statement controls how data points are represented on your plot. Three SYMBOL statement options are used here.

The INTERPOL= option specifies how data points are connected. There are several choices for connecting data points: INTERPOL=NONE leaves points unconnected; INTERPOL=JOIN draws a straight line between adjacent points. (See End Notes for more on the INTERPOL= option.)

The LINE= option defines the line style used to connect points. Type "1" is a solid line and is also the default if you do not specify a LINE= setting. Line types are listed in "Selecting Line Types" in Chapter 16, "The SYMBOL Statement," *SAS/GRAPH Software: Reference, Version 6, First Edition, Volume 1.*

VALUE= defines a character to be used to mark data points. VALUE=NONE means that individual data points will not be marked.

16 — The PROC statement begins the GPLOT procedure. The PHOSTEMP data set, created in the PROC SORT step (lines 1-3), is named for input.

18-19 — The TITLE statement defines the first line of the title for the PROC GPLOT output. (TITLE is equivalent to TITLE1.) The title text size is set to 5% of the graphics area, in the SWISS typeface. The text itself is in quotes on line 19.

20-21 — The TITLE2 statement defines text for the second title line. This second title line will be smaller than the first, 4% of the graphics area versus 5%. The typeface is SWISS, and the text is in quotes on line 21.

23 — The PLOT statement begins on this line with the keyword "PLOT" and ends on line 26 with the semicolon. The PLOT statement names the *x* and *y* variables to plot. The variable to the left of the * (asterisk) is the *y* or vertical axis variable. Fish count is shown on the *y* axis. Phosphorous concentration is shown on the *x* axis. PLOT statement options follow the "/".

24 — VAXIS= is a PLOT statement option. It applies a set of axis definitions to the vertical plot axis. VAXIS=AXIS1 assigns the AXIS1 definitions established by the AXIS statement on lines 7-8.

25 — The HAXIS= option applies the AXIS2 definitions to the horizontal plot axis. AXIS2 definitions are set by the AXIS statement on lines 10-12.

26 — The semicolon ends the PLOT statement that begins on line 23.

28 — The RUN statement executes the GPLOT procedure.

✳ Results ## Count Versus Concentration Plot

The completed plot appears below. Data points are displayed as specified by the SYMBOL statement (line 14). Points are connected by a line (INTERPOL=JOIN), but not marked (VALUE=NONE). Plot text is displayed according to the settings in the TITLE and AXIS statements.

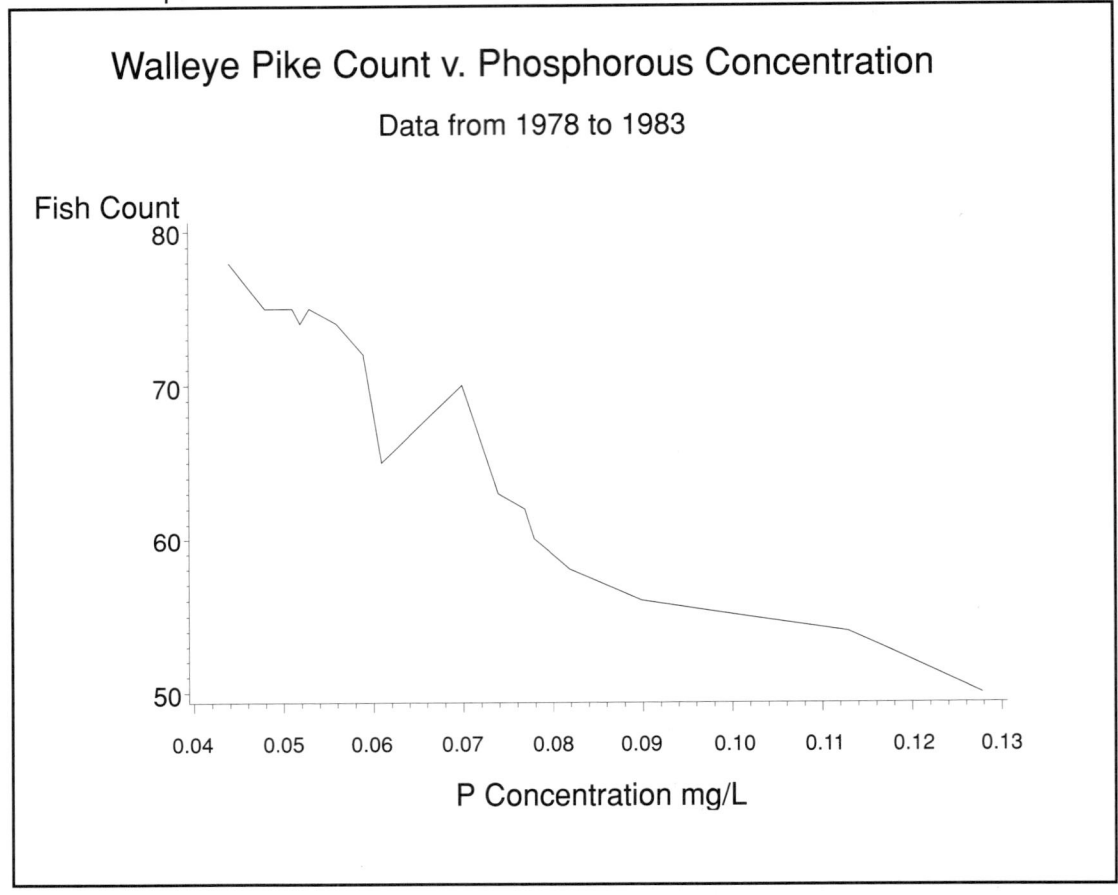

Fish count plot

Walleye Pike Count v. Phosphorous Concentration

Data from 1978 to 1983

[Line plot with Fish Count on y-axis (50 to 80) and P Concentration mg/L on x-axis (0.04 to 0.13), showing a generally decreasing trend from about 78 at 0.044 to about 51 at 0.13]

🔍 Closer Look Program 19 demonstrates a straightforward X-Y graph. **PROC GPLOT can also generate other plot types.** These include plots with multiple lines on a single set of axes, area fill plots, and graphs with separate left and right vertical axes. See Part 5, Two-Dimensional Plots" in *SAS/GRAPH Software: Usage, Version 6, First Edition* for examples.

214 Part 3 - Presenting Your Data

The SYMBOL Statement, Sorting, PROC PLOT

The SYMBOL statement. The SYMBOL statement helps you control the appearance of your plot. There are several SYMBOL statement options. LINE=, INTERPOL=, and VALUE= appear in Program 19. There are other options that control color, thickness of lines, size of plotting marks, etc.

The INTERPOL= option in the SYMBOL statement determines how the relationship between points will be presented. INTERPOL= settings include data smoothing, regression line fitting, and high-low plots.

In the modified version of Program 19 below, the SYMBOL1 statement is changed so that points are not connected by lines and each point is represented by a mark (VALUE=DOT). The resulting plot follows.

```
SYMBOL1 INTERPOL=NONE VALUE=DOT;

PROC GPLOT DATA=PHOSTEMP;
   TITLE HEIGHT=5 PCT FONT=SWISS
      'Walleye Pike Count v. Phosphorous Concentration';
   TITLE2 HEIGHT=4 PCT FONT=SWISS
      'Data from 1978 to 1993';
   PLOT COUNT*PHOS /
      VAXIS=AXIS1
      HAXIS=AXIS2
   ;
RUN;
```

The fish count plot showing points only

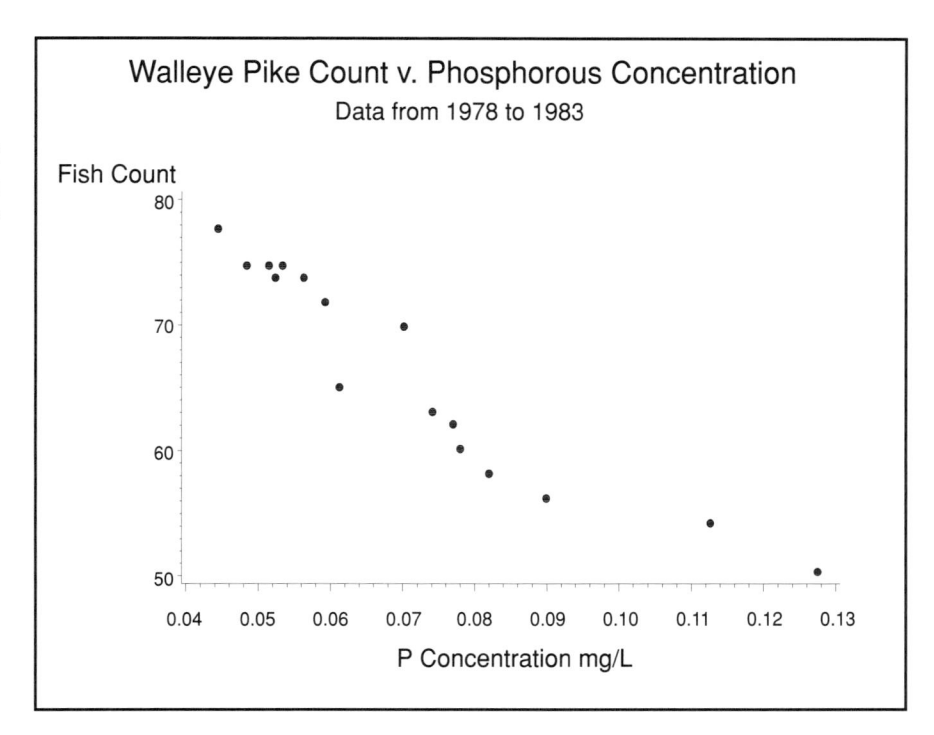

Walleye Pike Count v. Phosphorous Concentration
Data from 1978 to 1983

Chapter 19 - Creating Line Graphs and Plots

📝 **SYMBOL statement options.** SYMBOL options remain in effect until you explicitly change them. This can be confusing, especially when working interactively in display manager. For example, suppose SYMBOL1 definitions are set with the following statement:

```
SYMBOL1 INTERPOL=JOIN LINE=1 VALUE=NONE;
```

Points plotted using the SYMBOL1 settings will be connected by a line (INTERPOL=JOIN), but not marked (VALUE=NONE). If later in your SAS session you decide you do not want to connect points with a line, you could try the following SYMBOL statement:

```
SYMBOL1 INTERPOL=NONE;
```

The resulting plot would be blank because the VALUE=NONE option from the previous SYMBOL1 statement is still in effect. The INTERPOL=NONE setting means: *Don't connect the points.* The result is a plot with no points and no lines. To mark points you have to specify a VALUE= character such as VALUE=DOT. To reset all symbol definitions to their defaults, execute the statement: **GOPTIONS RESET=SYMBOL;**

📝 **Sorting plot data.** Why is the PROC SORT step necessary in Program 19? When you connect points with the SYMBOL statement's INTERPOL=JOIN option, points are joined in the order they are encountered in the input data set, not in order by the *x*-axis values. For the DNRFISH.PHOSTEST data set this means that the connecting line would begin with the point for the year 1978, then go to the point for 1979, etc. The following plot shows what happens if you plot the DNRFISH.PHOSTEST data set without sorting:

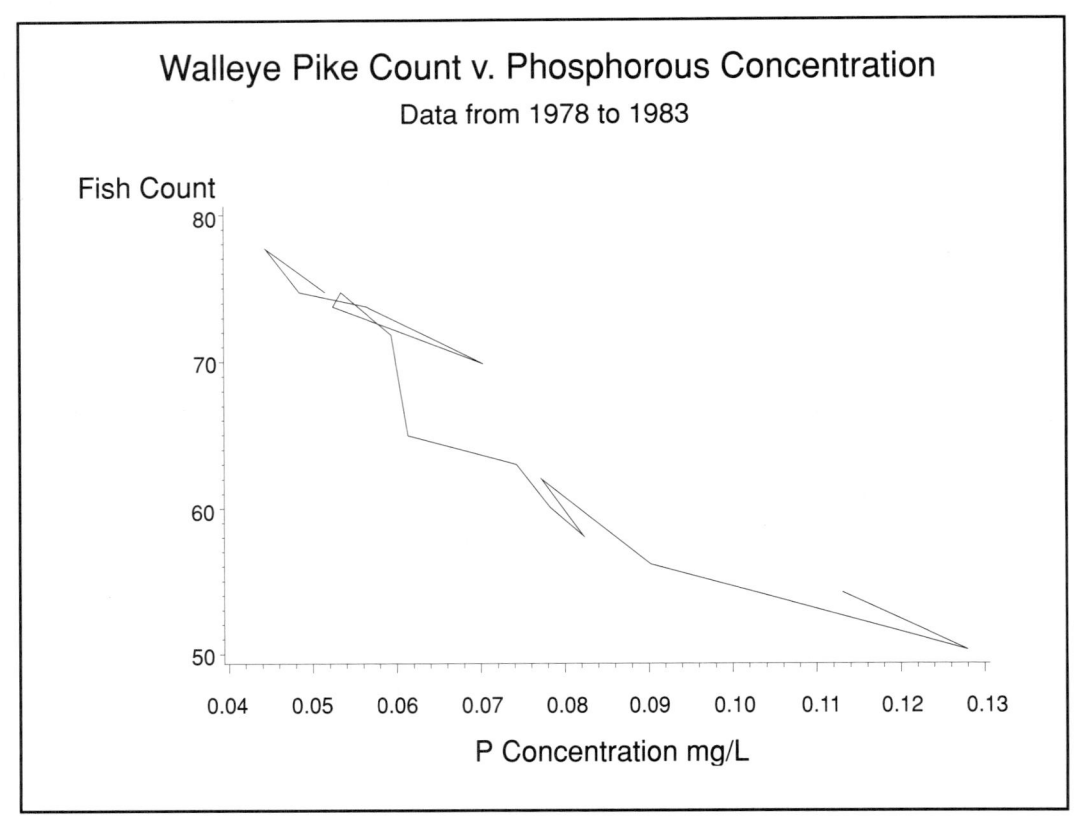

The fish count plot from the unsorted data set

Walleye Pike Count v. Phosphorous Concentration
Data from 1978 to 1983

216 Part 3 - Presenting Your Data

Character plots. If SAS/GRAPH software is not available on your system, you can use the PLOT procedure in base SAS software to create a character-based plot. Following the program is the PROC PLOT version of the fish count graph:

```
PROC SORT DATA=DNRFISH.PHOSTEST OUT=PHOSTEMP;
   BY PHOS;
RUN;
PROC PLOT DATA=PHOSTEMP;
   TITLE 'Walleye Pike Count v. Phosphorous Concentration';
   TITLE2 'Data from 1978 to 1993';
   PLOT COUNT*PHOS;
RUN;
```

The fish count graph using PROC PLOT

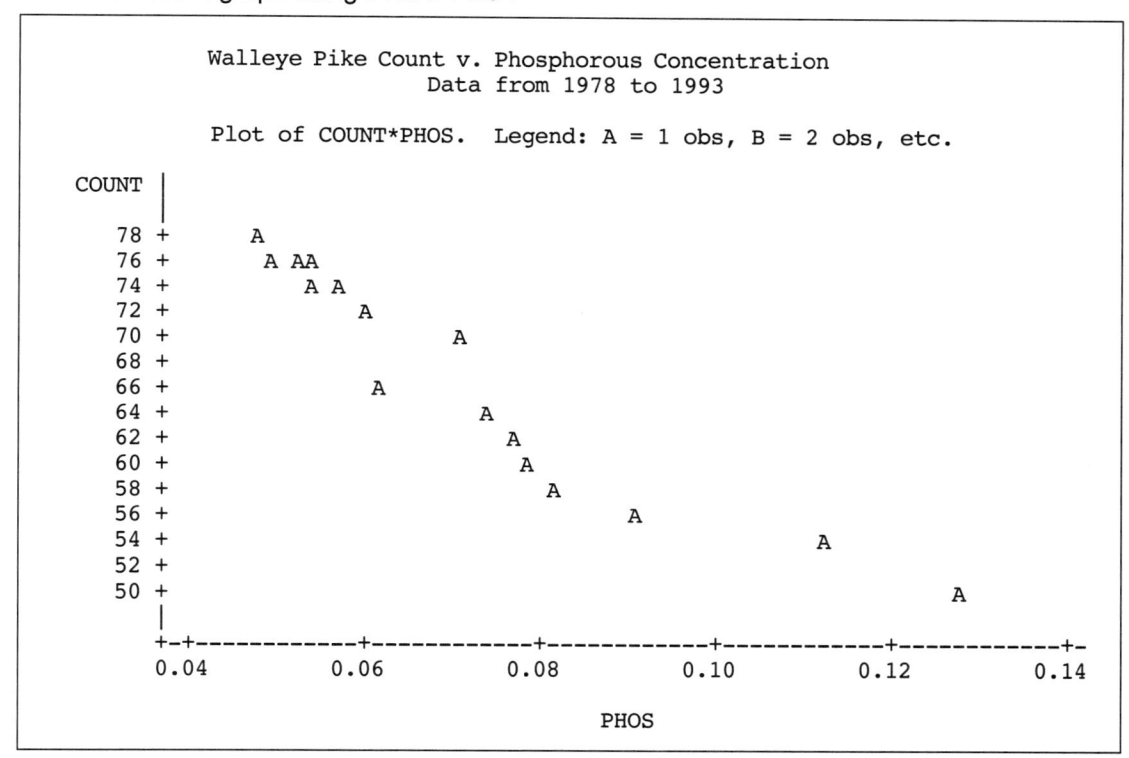

```
             Walleye Pike Count v. Phosphorous Concentration
                         Data from 1978 to 1993

         Plot of COUNT*PHOS.   Legend: A = 1 obs, B = 2 obs, etc.

   COUNT |
         |
     78 +        A
     76 +      A AA
     74 +       A A
     72 +        A
     70 +            A
     68 +
     66 +       A
     64 +          A
     62 +           A
     60 +           A
     58 +            A
     56 +              A
     54 +                   A
     52 +
     50 +                         A
         |
         +-+-------------+-------------+-------------+-------------+-------------+-
          0.04          0.06          0.08          0.10          0.12          0.14
                                            PHOS
```

Chapter 19 - Creating Line Graphs and Plots 217

Creating Line Graphs and Plots

1. Make sure you know the name and have access to the SAS data set with which you want to work. This may require a LIBNAME statement. (See Chapter 2 in this book for a LIBNAME example.)

2. Make sure you know the name of the graphics device driver where your chart will be printed or displayed. Name the output device with the DEVICE= option in a GOPTIONS statement. Use PROC GDEVICE to list drivers.

3. Use the SYMBOL statement to control how points are displayed and connected. To connect points with a line, use SYMBOL INTERPOL=JOIN.

4. If you are going to join points, sort the plot data by the *x*-axis variable to ensure that lines are connected correctly.

5. Use AXIS statements to control axis and tick mark labels.

6. Use PROC GPLOT. Name the variables you want to plot in a PLOT statement: PLOT *y-variable*x-variable*.

7. To apply axis settings, use the HAXIS= and VAXIS= options in the PLOT statement. PLOT statement options follow a "/".

PROC GPLOT, Graphics Controls

AXIS statement
Chapter 9, "The AXIS Statement," *SAS/GRAPH Software: Reference, Version 6, First Edition, Volume 1*

more plotting examples
Part 5, "Two-Dimensional Plots," *SAS/GRAPH Software: Usage, Version 6, First Edition*

PROC GPLOT general reference
Chapter 31, "The GPLOT Procedure," *SAS/GRAPH Software: Reference, Version 6, First Edition, Volume 2*

SAS/GRAPH fonts
Chapter 6, "SAS/GRAPH Fonts," *SAS/GRAPH Software: Reference, Volume 1*

SAS graphics options (GOPTIONS)
Chapter 5, "Graphics Options and Device Parameters Dictionary," *SAS/GRAPH Software: Reference, Volume 1*

SYMBOL statement
"Introduction to the SYMBOL Statement" in Chapter 19, "Introduction to Plots," *SAS/GRAPH Software: Usage;*
Chapter 16, "The SYMBOL Statement," *SAS/GRAPH Software: Reference, Volume 1*

Index

A

alternative output format for PRINT procedure 50
APPEND procedure 109
ascending sort order as default 49
assignment statements
   assigning text string to variables 185, 199
   changing date value, example 114
asterisk (*)
   See crossing operator (*)
at sign (@)
   indicating column location 26, 156
auto insurance claims report
   See custom reports, creating
averages, finding 73-80
   average crop yield in three counties 74-75
   average yield report 78
   averaging across all observations 76
   default output of MEANS procedure 78, 79
   program for determining averages 76-77
   summary 80
axis labels
   fish count plot example 212
   grouped bar chart example 200
   overriding default labeling 178
   stacked bar chart example 186, 189, 191
   vertical bar chart example 170
AXIS statements
   controlling axis and tick mark labels 191
   FONT= option 170, 177, 186, 200, 212
   global statement 170, 186, 187
   HEIGHT= option 170, 177, 186, 200, 212
   LABEL= option 170, 186, 191, 203, 204, 212
   maximum number of axis definitions 170, 186
   ORDER= option 172
   VALUE= option 186, 191, 203, 204, 212
AXIS1 statement
   assigning AXIS1 settings to response axis 171, 172
   fish count plot example 212
   grouped bar chart example 200, 202
   stacked bar chart example 186, 189
   vertical bar chart example 170
AXIS2 statement
   assigning AXIS2 settings to midpoint axis 171, 172
   fish count plot example 212
   grouped bar chart example 200, 202
   stacked bar chart example 186, 189
   vertical bar chart example 170
AXIS3 statement 200, 202

B

Backward key 39
bar charts
   See grouped bar charts, creating
   See stacked bar charts, creating
   See vertical bar charts, creating
base SAS software 5
blank space concatenation operator, TABULATE procedure 145
blanks
   embedded blanks in data 18
   separating fields in data 18
book inventory example
   See unique values, finding
BY groups 97, 158
BY statement
   alternative output format for PRINT procedure 50
   auto insurance claims report, example 154
   creation of FIRST. and LAST. variables 122, 123, 125, 154
   DESCENDING option 154
   effect on PRINT procedure 47
   ensuring that data is sorted in BY-variable order 49
   generating frequency counts 60
   grouping data by formatted values 94, 96-98
   grouping data by region, example 46
   inability to read external files 155
   NOTSORTED option 98
   processing grouped data without sorting 98
   required with SORT procedure 46, 212
   using with FORMAT procedure and FREQ procedure 94, 96
   using with SET statement 122, 123
BY variables 46, 47, 49

C

CARDS statement 16
CENTER system option 27
character-based charts and plots
   character plots 217
   vertical bar charts 173
character formats
   See formats

character variables
   See variables
CHART procedure 173
CLASS statement, MEANS procedure
   averaging data for one variable 77, 78
   grouping data 65, 66
   grouping data by date 84, 85
   omitting when averaging across all observations 76
   order of variables 57
CLASS statement, TABULATE procedure 143
colon (:)
   ending labels 156
color list, default 201
COLOR= option, PATTERN statement
   BLACK value 186
   specifying color for fill areas 177
combining data sets
   See SAS data sets, concatenating
comma operator (,)
   TABLE statement 145, 147
concatenating data sets
   See SAS data sets, concatenating
CONTENTS procedure 116
continuous variables, counting 60
correcting variables
   See observations, editing with DATA step
   See observations, editing with FSEDIT procedure
counting occurrences
   See frequency analysis
crop yield report
   See averages, finding
crossing operator (*)
   example 145
   interpretation of 146
   purpose and use 143
   table output with crossing operator removed 146
custom reports, creating 149-165
   alternatives to DATA step report writing 162
   claims report layout 150
   DATA step report writing 152-157
   dollar signs in reports 161
   features of DATA step report writing 160
   generating auto insurance claims report 150-151
   layout claims report, example 158-159
   page settings 161
   planning worksheet 160
   PRINT procedure output, example 163
   summary 164
customer payment records
   See observations, editing with FSEDIT procedure

D

data, editing
   See observations, editing with DATA step
   See observations, editing with FSEDIT procedure
data, grouping by dates 81-89
   displaying dates and days 87-88
   finding monthly total rainfall 82
   grouping by weekday names 87-88
   monthly precipitation report 86
   program for grouping by months 84-85
   summary 89
   using FORMAT statement and MEANS procedure 83-86
data, grouping by formatted values 91-100
   character formats 98
   comparing math scores 92-93
   creating custom format 94-96
   grouping unsorted data 98
   lookup tables 99
   program for grouping math scores 94-96
   summary 100
   test score report 97
data, sorting and grouping 43-52
   alternative output format for PRINT procedure 50
   creating sales report grouped by region 44-48
   grouping data 49
   grouping data with CLASS statement 65, 66
   preserving original data set 49
   program for sorting 45-46
   reasons for sorting, illustration 49
   regional sales report, example 47-48
   sorting data 49
   summary 52
   using formats 51
data entry 13-20
   See also external files, reading
   blanks embedded in data 18
   handwritten form, example 14
   input styles 18
   instream data 18
   program including instream data 15-16
   sales activity list, example 17
   summary 19

using FSEDIT procedure 37-38
using instream data, example 15-16
DATA= option
   FREQ procedure 57
   PRINT procedure 46
   SORT procedure 46
data set options
   DROP= 105
   IN= 104, 106
   KEEP= 105
   OBS= 105
   RENAME= 105
data sets
   See SAS data sets
DATA step functions
   capabilities 70
   SUM 70
DATA steps
   calculating percentages 135
   compared with other programming languages 6
   concatenating data sets 104-106
   creating new data set 40
   entering data, example 15-16
   examples 7
   finding and correcting observations 113-114
   finding unique values 122, 123
   implicit instructions added by SAS System 114
   __NULL__ keyword 154
   preserving input data set while writing output data set 116
   processing data with subsetting IF statement 126
   purpose and use 6-7
   reading external files 23, 25, 27
   replacing data sets 116
   structuring data for grouped bar charts 197, 199-200
   structuring data for stacked bar charts 183-186, 190
   summing variables within an observation 70
   writing custom reports 152-160
data validity checking 18
date formats
   DATE7. 25, 26, 27, 44, 82, 112, 150
   MMDDYY$w.$ 51
   MMDDYY8. 32, 38, 156
   MONNAME$w.$ 87
   MONTH$w.$ 87
   MONYY$w.$ 51
   MONYY5. 83, 84, 85
   QTR$w.$ 87
   special class of numeric formats 51
   WEEKDATE$w.$ 87-88
   WEEKDAY$w.$ 87
   YEAR$w.$ 87
   YYMMDD$w.$ 51
date informats
   MMDDYY8. 24, 26, 41
   validating data entry 41
DATE system option 27
date values
   See also date formats
   changing 113, 114
   controlling data entry with informats 41
   handling date values 24
   stored as numeric variables 24, 83
dates, grouping data by
   See data, grouping by dates
DATE7. format
   assigning to date variable 25, 26
   examples 27, 44, 82, 112, 150
day-to-month grouping
   See data, grouping by dates
days of the week
   See WEEKDATE$w.$ format
decimal places
   FREQ procedure 133
   specifying in formats 51
deleting variables
   See observations, editing with FSEDIT procedure
DESCENDING option, BY statement 154
DEVICE= option, GOPTIONS statement
   checking or changing from OPTIONS window 170
   specifying graphics output device 170, 186, 200, 212
DISCRETE option
   chart without DISCRETE option, example 175
   purpose 175, 178
   used only with numeric chart variables 188
   vertical bar chart example 171
discrete variables, counting 60
display manager 8-9
DO groups
   executing statements when IF condition is true 117
   printing headers in custom report 155
   syntax 117
   terminating with END statement 155

dollar sign ($)
    indicating character formats 51, 98
    indicating character variables 16
    specifying character data in informats 26
    $w. format 51
DOLLAR5. format 187, 189, 201
DOLLAR7. format 51
DOLLAR11.2 format 46, 68
DOLLAR13.2 format 161
DOUBLE option, PRINT procedure 47
DROP= data set option 105

E

editing observations
    See observations, editing with DATA step
    See observations, editing with FSEDIT
        procedure
ELSE statement
    See IF-THEN/ELSE statements
END statement
    terminating DO groups 155, 156
    terminating SELECT group 106, 107
entering data
    See data entry
equal sign (=)
    absence of indicating sum statements 155
equal sign modifier (=)
    TABULATE procedure 143, 145
equipment purchases report, generating
    See sums, finding
experimental results table
    See tables, creating
external files 23, 28
external files, reading 21-30
    creating patient visit report 22-27
    DATA step 155
    enclosing filename in quotes 25
    patient visit file, illustration 22
    patient visit report, example 27
    process for reading 28
    program for reading 24-26
    steps for reading 23
    summary 29

F

FILE statement 154
FILENAME statement
    assigning fileref to external file 25, 27
    specifying external files 23
    syntax 25

filerefs
    assigning to external file 25
    definition of 25
    preferred instead of specifying filenames
        directly 25
files, reading
    See external files, reading
FIND command 41, 42
finding variables for editing
    See observations, editing with DATA step
    See observations, editing with FSEDIT
        procedure
FIRST. and LAST. variables
    creating with BY and SET statements 122
    finding unique values 122-123, 127
    grouping categories in custom report 154-155
    inability to read external files 155
    purpose and use 125
fish count plot
    See line graphs and plots, creating
fixing shipping department log
    See observations, editing with DATA step
FONT= option
    AXIS statements 170, 177, 186, 200, 212
    FOOTNOTE statement 201
    SWISS value 170, 186, 200, 212
    TITLE statement 171, 177
    ZAPF value 177, 201
fonts, proportionally spaced 161
FOOTNOTE statement
    compared with NOTE statement 204
    grouped bar chart example 201
FORMAT procedure 94-96, 98
    creating custom formats 94-96
    creating lookup tables 99
    uses 98
    using with BY statement 94
    using with FREQ procedure 94, 96
FORMAT statement 83-85
    creating custom formats 96
    displaying dollar amounts 46, 47, 68
    forcing procedures to process formatted values
        83-85
    not available for MEANS procedure statistics
        67
    specifying formats 25, 187, 189, 201
    syntax 46, 84
formats
    character formats 51, 98
    components of date formats 85
    creating custom formats 94-96, 98

Index 224

date formats 51, 85, 87
DATE7. 25, 26, 27, 44, 82, 112, 150
DOLLAR5. 187, 189, 201
DOLLAR7. 51
DOLLAR11.2 46, 68
DOLLAR13.2 161
form of formats, example 84
generating lookup tables 99
$HEXw. 51
MMDDYYw. 51
MMDDYY8. 32, 38, 156
MONNAMEw. 87
MONTHw. 87
MONYYw. 51
MONYY5. 83, 84, 85
numeric formats 51
PERCENTw.d 51
permanent formats 96
purpose and use 26
QTRw. 87
temporary formats 96
$VARYINGw. 51
$w. 51
WEEKDATEw. 87-88
WEEKDAYw. 87
width and decimal places (w.d) 51
width of date formats 85
WORDSw. 51
YEARw. 87
YYMMDDw. 51
Zw.d 51
$ preceding character formats 51, 98
4.2 142
formatted input
  reading data with embedded blanks 18
  reading external files 26
formatted PUT statement 156
formatted values, processing
  See data, grouping by dates
  See data, grouping by formatted values
Forward key 39
FREQ procedure 57-62, 131-136
  calculating percentages 131-136
  compared with MEANS procedure 69, 135
  control of decimal places not available 133
  counting occurrences 57-58
  cumulative data not included in output data set 61
  DATA= option 57
  generating output data sets 61
  options for controlling output 59
  ORDER= option 59, 132
  sample program 57-58
  specifying reporting order 59
  statistics produced in reports 59, 69, 97
  TABLE statement 58, 59, 96, 131-132, 135
  using with BY statement and FORMAT statement 94, 96-97
  WEIGHT statement 131-132, 134
frequency analysis 55-62
  continuous variables 60
  creating quality control report 56-59
  cumulative frequency 59
  cumulative percent 59
  discrete variables 60
  generating data set with frequency information 61
  options for suppressing cumulative and percentage data 59
  other methods of generating 47
  percent 59
  procedures available for frequency analysis 60
  program for frequency analysis 57-58
  program using FREQ procedure 57-58
  rejected parts report 59
  summary 62
FSEDIT procedure 33-39
  capabilities 33
  correcting variables 37
  data validity checking 18
  editing observations 33-39
  exiting FSEDIT session 39
  finding variables 34-36, 41
  requirements for using 42
  scrolling through observations 39
  using menus 33
  windowing interface, illustration 9

G

GAXIS= option, VBAR statement
  assigning axis definitions 204
  grouped bar chart example 200, 202, 204
GCHART procedure
  chart using default text settings, example 192
  compared with CHART procedure 173
  grouped bar charts, creating 197-198, 201
  PATTERN statement 170
  RUN statement behavior 169
  stacked bar charts, creating 183-184, 187-188, 192
  vertical bar charts, creating 169-171
GDEVICE procedure 170

global settings, resetting 187
global statements
 AXIS 170, 186, 187
 characteristics 58
 definition of 6
 effects on fill patterns 187
 GOPTIONS 170
 LEGEND 187
 PATTERN 170, 187
 SAS/GRAPH global statements not supported by
  CHART procedure 173
 TITLE statement 58
GOPTIONS statement
 DEVICE= option 170, 186, 200, 212
 global statement 170
 RESET= option 187, 216
 ROTATE= option 170, 186, 200, 212
 TARGETDEVICE= option 212
GPLOT procedure 211, 213
graphics options
 See GOPTIONS statement
graphics output devices
 See also DEVICE= option, GOPTIONS statement
 obtaining list of drivers 170
 SAS/GRAPH software support 170
group axis
 assigning axis definitions 204
 grouped bar chart example 200, 202
GROUP= option, VBAR statement
 combining with SUBGROUP= option 205-206
 grouped bar chart example 197, 198, 202, 203, 205-207
grouped bar charts, creating 195-208
 controlling chart appearance 204-206
 creating quarterly profits chart 196
 group AXIS settings 204
 placing text with NOTE statement 204
 program for creating 197-202
 quarterly profits chart, example 203
 simplifying the program 198
 structuring of data 197, 199-200
 summary 207
 using GROUP= or SUBGROUP= options 205-206
grouping data
 See data, grouping by dates
 See data, grouping by formatted values
 See data, sorting and grouping
growth rate table
 See tables, creating

H

handwritten data
 See data entry
HAXIS= option, PLOT statement 212, 213
HEIGHT= option, AXIS statements
 controlling text height 177
 fish count plot example 212
 grouped bar chart example 200
 stacked bar chart example 186
 units of measurement 177
 vertical bar chart example 170
HEIGHT= option, TITLE statement 171, 177
$HEXw. format 51
hourly utilization chart
 See vertical bar charts, creating

I

ID statement 50
IF statement, subsetting
 See subsetting IF statement
IF-THEN/ELSE statements 117
 DO groups 117
 finding variables for correction 113, 114
 syntax 117
IN= data set option 104, 106
INFILE statement
 reading external files 23, 25, 27
 specifying filenames directly 25
informats
 assigning to variables 41
 date informats 24, 26
 default numeric informat 26
 dollar sign ($), specifying character data 26
 form of informats, example 26
 MMDDYY8. 24, 26, 41
 purpose and use 26
 specifying in INPUT statement 26
 validating entry of dates 41
INPUT statement
 reading external files 23, 27
 reading external files, example 26
 reading instream data 16
 spanning multiple lines 26
input styles
 formatted input 18, 26
 list input 18
instream data
 definition of 15, 18
 limitations 18

interactive editing
    See observations, editing with FSEDIT procedure
INTERPOL= option, SYMBOL statement
    displaying relationship between points 215
    JOIN value 213, 214
    NONE value 213, 216

K

KEEP= data set option 105
KEEP statement 185, 199
key assignments 39

L

LABEL= option, AXIS statements
    assigning axis definitions 204
    fish count plot example 212
    grouped bar chart example 203
    stacked bar chart example 186, 191
    vertical bar chart example 170
LABEL= option, LEGEND statement 187, 191
LABEL option, PRINT procedure 47
LABEL statement 142
labels
    See axis labels
    See statement labels
    See tick mark labels
landscape orientation 170
LAST. *variable*
    See FIRST. and LAST. variables
LEGEND= option, VBAR statement
    applying LEGEND statement definitions 188, 189
    grouped bar chart example 205
    using with SUBGROUP= option 191
LEGEND statement 191
    controlling appearance of legends 187, 191
    global statement 187
    grouped bar chart example 205
    LABEL= option 187, 191
    maximum number of legend definitions 187
    stacked bar chart example 189, 191
    VALUE= option 187, 191
LENGTH statement 16
LIBNAME statement 25
librefs
    definition of 25
    first part of two-part names 25, 32
    SALEDEPT example 46
    specifying for external file 25
line graphs and plots, creating 209-219

checking output on display 212
count versus concentration plot, example 214
creating character-based plot 217
creating fish count plot 210
order of joining points 216
program for creating 211-213
representing points in plots 215-216
sorting plot data 216
summary 218
using SYMBOL statement 215-216
LINE= option, SYMBOL statement 213, 215, 216
LINESIZE= system option 161
LINESLEFT= option 154, 161
LINK statement 155, 156, 157
list input 18
LL= option
    See LINESLEFT= option
LOCATE command 41, 42
LOG window 10
lookup tables
    generating with custom formats 99
low inventory report example
    See unique values, finding

M

math scores, comparing
    See data, grouping by formatted values
MAXDEC= option, MEANS procedure 77, 84
MAXIS= option, VBAR statement
    grouped bar chart example 200, 202
    stacked bar chart example 186, 188, 189
    vertical bar chart example 171, 172
MEAN statistic
    MEANS procedure 77
    TABULATE procedure 143
MEANS procedure 65-71, 76-79, 83-86
    See also sums, finding
    average county yields report, example 78
    averaging across all observations 76
    calculating percentages 135
    CLASS statement 65, 66, 67
    compared with FREQ procedure 69, 135
    counting occurrences 69
    default output 79
    default statistics 79
    equipment purchases report, example 67
    finding averages 76-79
    format limitations in reports 67
    frequency analysis 60
    generating output data set 67, 68-69
    grouping data by formatted date values 83-86

limitations for summing variables 70
MAXDEC= option 77, 84
MEAN statistic 77
N statistic 69
NOPRINT option 69
ORDER=FORMATTED option 88
OUTPUT statement 69
output statistics not affected by FORMAT statement 67
statistics available 65, 79
subtitle added to reports 67, 78, 86
SUM statistic 66, 84
suppressing printed output 69
variables in output data set 69
midpoints axis
bar chart example 171
grouped bar chart example 200, 202
stacked bar chart example 188
MIDPOINTS= option, VBAR statement 172, 188
MMDDYY*w.* format 51
MMDDYY8. format 32, 38, 156
MMDDYY8. informat 24, 26, 41
MONNAME*w.* format 87
months, grouping data by
See data, grouping by dates
MONTH*w.* format 87
MONYY*w.* format 51
MONYY5. format 83, 84, 85

N

N statistic
MEANS procedure 69
PRINT procedure 47, 60
TABULATE procedure 146
NAME command 41, 42
naming conventions
two-part names for data sets 25, 32
variables 26
NOCARDIMAGE system option 18
NOCUM option, TABLE statement 135
NOOBS option, PRINT procedure 47
NOPRINT option
MEANS procedure 69
TABLE statement 61
NOREPLACE system option 116
NOTE statement 204
NOTSORTED option, BY statement 98
NULL keyword 154
NUMBER system option 27
numeric formats
See formats 51

numeric variables
compared with character variables 113
date values stored as numeric variables 24
default SAS System variable 16

O

OBS= data set option 105
observations, editing with DATA step 111-118
corrected shipments log, example 115
correcting date value 113
fixing shipping department log 112
preserving input SAS data sets 116
program for correcting data 113-114
replacing SAS data sets 116
summary 118
using IF-THEN/ELSE statements 117
observations, editing with FSEDIT procedure 31-42
adding new record 37-38
correcting variables 37
deleting observations 39, 40
entering dates 41
finding observations 34-36, 41
finding observations by number 41
summary 42
updated PAYMENTS data set, example 40
updating customer payment records 32-42
using FSEDIT procedure 33-39
OPTIONS window 170
ORDER= option
AXIS statements 172
FREQ procedure 59, 132
MEANS procedure 88
OTHERWISE statement 107
OUT= option
OUTPUT statement 69
SORT procedure 49, 212
OUTPUT statement, DATA step
data set for grouped bar charts, example 199-200
data set for stacked bar charts, example 185, 186
OUTPUT statement, MEANS procedure
alternative form for specifying statistics 69
generating SAS data set 69
inheriting variable names in output data set 69
OUT= option 69
OUTPUT window 154

P

__PAGE__ specification, PUT statement 155, 156
PAGEBY statement 47
PAGESIZE= system option 161
parentheses ()
   grouping LABEL= and VALUE= settings 186
   parentheses associative operator, TABULATE procedure 145
park visits, calculating
   See percentages, finding
patient visit report
   See external files, reading
PATTERN statements
   COLOR= option 177, 186
   default color list 201
   fill values and colors 177, 201
   global statement 170, 187
   VALUE= option 170, 177, 186, 189, 201
PATTERN1 statement
   grouped bar chart, example 201, 203
   stacked bar chart, example 186
PATTERN2 statement 186
percentages, finding 129-136
   calculating park visits by month 130
   DATA step language 135
   FREQ procedure with WEIGHT statement 131-132, 134-136
   MEANS procedure 135
   monthly percentage report 133
   program for finding percentages 132
   summary 136
   TABULATE procedure 135
PERCENT$w.d$ format 51
period (.)
   following formats and informats 26
permanent data sets 16, 32
PLOT procedure 217
PLOT statement
   fish count plot example 213
   HAXIS= option 212, 213
   slash (/) preceding options 213
   VAXIS= option 212, 213
plots
   See line graphs and plots, creating
portrait orientation 170
PRINT file 154
PRINT procedure
   alternative output format 50
   compared with DATA step report generation 162-163
   concatenated data set example 107, 108
   corrected shipping log, example 114, 115
   DATA= option 46
   effect of BY statement 47
   generating frequency counts 60
   list of last books, example 123, 124
   N statistic 47, 60
   options for controlling page layout 47
   patient visit report, example 26
   printing data set generated by MEANS procedure 68
   sales activity list, example 15, 16, 17
   updated observations, example 40
procedures
   See SAS procedures
PROGRAM EDITOR window 9, 169
PUT statement
   formatted PUT statement 156
   output filename 154
   __PAGE__ specification 155, 156
   printing summary information in reports 157
   writing headers in reports 157

Q

QTR$w.$ format 87
quality control report
   See frequency analysis
quarterly profits chart
   See grouped bar charts, creating
quotation marks
   enclosing character literals 114
   enclosing external filenames 25

R

rainfall totals, grouping
   See data, grouping by dates
RAXIS= option, VBAR statement
   grouped bar chart example 200, 202
   stacked bar chart example 186, 188, 189
   vertical bar chart example 171, 172
reading files
   See external files, reading
regional sales report example
   See data, sorting and grouping
RENAME= data set option 105
REPORT procedure 162
reports
   See also custom reports, creating
   See also tables, creating
   average yield report 78, 79
   equipment purchases report 67
   list of last books 124

monthly percentage report 133
monthly precipitation report 86
patient visit report 27
regional sales report, example 47-48, 50
rejected parts report 59
sales activity list, example 17
system options for controlling page layout 27
test score report 97
RESET= option, GOPTIONS statement 187, 216
response axis
bar chart example 171
grouped bar chart example 200, 202
stacked bar chart example 188
RETURN statement
custom report example 156, 157
multiple uses 157
ROTATE= option, GOPTIONS statement 170, 186, 200, 212
ROUND option, PRINT procedure 47
RUN statement
behavior when running GCHART procedure interactively 169
RUN statement examples
DATA steps 16, 26, 107, 114, 123, 157, 186, 200
FORMAT procedure 96
FREQ procedure 58, 96, 132
GCHART procedure 171, 188, 202
GPLOT procedure 213
MEANS procedure 66, 77, 85
PRINT procedure 16, 46, 107, 114, 123
SORT procedure 46, 96, 154, 212
TABULATE procedure 143

S

sales report grouped by region
See data, sorting and grouping
sample SAS data sets
ACCOUNT.EQUIPMNT 64
ACCOUNT.EQUIPSUM 68
ACCOUNT.QTRPROF 196
AGDEPT.YIELDS 74-75
AGLAB.GROW1 141
AUTODIV.CLAIMS 151
DNRFISH.PHOSTEST 210
EXPENSES.UTILITY 182, 183
PAYMENTS.JULAUG 32
PERFORM.CAPUTIL 168
PROFITS 70
QCDATA.LINECNT 61
QCDATA.REJECTS 56

REDCLIFF.VISITS 130
RESULTS.ALLEXPER 108
RESULTS.EX91102 102
RESULTS.EX91103 103
RESULTS.EX91104 103
SALEDEPT.SALES 44
SCORES.MATH8 92-93
SDEPT.SHIPLOG 112, 115
SLEUTH.BOOKS 120-121
UTLTEMP 190
WEATHER.PRECIP 82
SAS data sets
See also sample SAS data sets
creating new sorted output data set 49
definition of 7
generating with FREQ procedure 61
generating with MEANS procedure 67, 68-69
inheriting variable names in output data set 69
organization, illustration 7
permanent data sets 16, 32
preserving input data set while writing output data set 116
removal of deleted observations 40
replacing with changed version 116
temporary data sets 16
two-part names 25, 32
SAS data sets, concatenating 101-110
adding variable to distinguish observations 104, 106
combined data set, example 108
combining experimental results 102-103
program for concatenating 104-107
summary 110
using APPEND procedure 109
SAS Display Manager System 8-9
SAS/FSP software
See also FSEDIT procedure
required for FSEDIT procedure 33
SAS/GRAPH software
advantages 173
support for graphics output devices 170
SAS language
implicit instructions added to DATA steps 114
SAS log 10
SAS procedures
PROC steps 6
PROC steps, examples 7
purpose 5

SAS programs 6-7
    DATA steps 6
    examples 7
    PROC steps 6
    rules for writing programs 8
    steps, illustration 6
SAS software 5
SAS System 5
SAS system options
    CENTER 27
    controlling page layout 27
    DATE 27
    LINESIZE= 161
    NOCARDIMAGE 18
    NOREPLACE 116
    NUMBER 27
    PAGESIZE= 161
SEARCH command 41, 42
searching for variables
    See observations, editing with DATA step
    See observations, editing with FSEDIT
        procedure
SELECT group 106, 107
SELECT statement
    adding variable while concatenating data sets
        104, 106
    using as alternative to IF-THEN/ELSE
        statements 117
    using expressions 107
selecting and changing data
    See observations, editing with DATA step
    See observations, editing with FSEDIT
        procedure
semicolon
    ending SAS statements 8
    terminating instream data 16
SET statement 109
    concatenating data sets 104, 106
    creating new data sets 40
    creation of FIRST. and LAST. variables 122,
        123, 154
    purpose and use 109
    reading data sets 113, 114, 154, 185, 199
    replacing data sets 116
    using with BY statement 122, 123, 154
shipment log, fixing
    See observations, editing with DATA step
slash (/)
    preceding options in PLOT statement 213
    preceding options in VBAR statement 171, 178,
        187, 202
    preceding TABLE statement options 61, 135

printing blank lines in reports 157
SORT procedure
    ascending order as default 49
    auto insurance claims report, example 154
    BY statement required 46, 212
    creating new sorted output data set 49
    DATA= option 46
    fish count plot example 212, 216
    OUT= option 49, 212
    sales report example 46
    test scores example 94, 96
sorting data
    See data, sorting and grouping
SPLIT= option, PRINT procedure 47
stacked bar charts, creating 181-194
    controlling appearance of axis and tick mark
        labels 191
    controlling legend settings 191
    creating utility expense chart 182
    data organization 190
    default text settings 192
    process for structuring data, illustration 183
    program for creating 184-188
    summary 193
    utility expense chart 189
statement labels 156, 157
statistics, generating
    See FREQ procedure
    See frequency analysis
    See MEANS procedure
STRING command 41, 42
SUBGROUP= option, VBAR statement
    combining with GROUP= option 205-206
    creating stacked bar charts 183, 187
    data structure requirements 190
    TYPE value 189
    using with LEGEND= option 191
subsetting IF statement 126
    checking for lines remaining on page 156
    compared with WHERE statement 126-127
    finding unique values 122-123
    implied THEN and ELSE actions 126
    purpose and use 126
    testing of FIRST. and LAST. variables 122,
        155, 156
SUM function 70
SUM option, MEANS procedure 66, 84
SUM= option, OUTPUT statement 69
SUM statement, in PROC PRINT
    patient visit report, example 26, 27
    regional sales report example 46, 47
    sales total example 15, 16, 17

231 Index

sum statements, in the DATA step 155
SUMMARY procedure 60
sums, finding 63-71
   See also MEANS procedure
   counting with MEANS procedure 69
   equipment purchases report 67
   generating equipment purchases report 64-67
   outputting summarized data into data set 68-69
   program for producing sum statistic 66
   sum statistic 65
   summary 71
   summing variables within an observation 70
SUMVAR= option, VBAR statement
   bar chart example 171
   chart created without SUMVAR= option,
      example 174
   combinations of SUMVAR= and TYPE= options,
      table 176
   data structure requirements 190
   grouped bar chart example 202, 203
   purpose 174
   stacked bar chart example 188
SYMBOL statement
   controlling appearance of plot 215
   fish count plot example 213, 215
   interactive effect of options 216
   INTERPOL= option 213, 214, 215, 216
   LINE= option 213, 215, 216
   resetting symbol definitions to defaults 216
   VALUE= option 213, 214, 215, 216

T

Table Producing Language (TPL) 145
TABLE statement, FREQ procedure
   calculating percentages 131-132
   generating output data sets 61
   NOCUM option 135
   NOPRINT option 61
   options for suppressing cumulative and
      percentage data 59
   OUT= option 61
   preceding options with slash (/) 61, 135
   quality control report example 58
   specifying variable for analysis 96
   suppressing cumulative statistics 135
TABLE statement, TABULATE procedure
   comma operator (,) 145, 147
   defining cell layout and content 143
   requirements for listing variables 143
   syntax 144
   Table Producing Language 145

tables, creating 139-148
   average growth rate table 144
   creating experimental results table 140-141
   crossing operator (*) 143, 145, 146
   dimensions of tables 147
   program for creating 142-143
   summary 148
   Table Producing Language 145
   using TABULATE procedure 142-148
TABULATE procedure 142-148
   average growth rate table, example 144
   blank space concatenation operator 145
   calculating percentages 135
   crossing operator (*) 143, 145, 146
   dimensions of tables 147
   equal sign modifier (=) 143, 145
   example program 142-143
   frequency analysis 60
   MEAN statistic 143
   N statistic 146
   parentheses associative operator 145
   statistics available 143
   Table Producing Language 145
   TABLE statement 143
   4.2 format 142
TARGETDEVICE= graphics option 212
temporary data sets 16
text settings, default 192
tick mark labels
   controlling with VALUE= option 186, 191
   fish count plot example 212
   grouped bar chart example 200
   stacked bar chart example 189
TITLE statement
   compared with NOTE statement 204
   equivalent to TITLE1 statement 201
   FONT= option 177
   global characteristics 58
   HEIGHT= option 177
TITLE statement examples
   auto insurance claims report 154
   average crop yield 77
   book inventory 124
   calculating percentages 132
   concatenated data set 108
   corrected shipping log 114
   creating tables 142
   equipment purchases report 66
   fish count plot 213
   grouped bar chart 201
   grouping test scores 96

monthly precipitation report 84
patient visit report 26
quality control report 58
sales activity report 16, 46
stacked bar chart 187
vertical bar chart 171
TITLE2 statement examples
   fish count plot 213
   grouped bar chart 201
TPL
   See Table Producing Language (TPL)
two-part names 25, 32
TYPE= option, VBAR statement 176
   combinations of SUMVAR= and TYPE= options, table 176
   statistics available 176

U

UNIFORM option, PRINT procedure 47
unique values, finding 119-128
   FIRST. and LAST. variables 122-123, 125, 127
   generating low inventory report 120
   list of last books 124
   program for finding one-and-only-one values 122-123
   subsetting IF statement 122-123, 126-128
   summary 128
utility expense chart example
   See stacked bar charts, creating

V

validating data
   See data validity checking
VALUE= option, AXIS statements
   assigning axis definitions 204
   fish count plot example 212
   grouped bar chart example 203
   stacked bar chart example 186, 191
VALUE= option, LEGEND statement 187, 191
VALUE= option, PATTERN statement
   EMPTY value 186, 189
   SOLID value 170, 186, 189, 201
   specifying fill patterns 177
VALUE= option, SYMBOL statement
   DOT value 216
   NONE value 213, 214, 216
   specifying representation of plot values 215
VALUE statement
   creating custom formats 95
   creating lookup tables 99
VAR statement
   creating tables, example 143
   finding averages, example 77, 78
   finding sums, example 66
   grouping data by dates, example 85
   inheriting variable names in output data set 69
   specifying variables for printing 46
variables
   See also FIRST. and LAST. variables
   assigning informats 41
   averaging across all observations 76
   BY variables 46, 47
   character variables compared with numeric variables 113
   continuous variables 60
   correcting 37
   date values stored as numeric variables 24
   defining length 16
   discrete variables 60
   enclosing character literals in quotes 114
   finding with FSEDIT procedure 34-36, 41
   forcing procedures to process formatted values 83-85
   inheriting variable names in output data set 69
   naming conventions 26
   numeric variables as default 16
   specifying for printing 46
   summing within an observation 70
$VARYINGw. format 51
VAXIS= option, PLOT statement 212, 213
VBAR statement
   DISCRETE option 171, 175
   GAXIS= option 200, 202, 204
   GROUP= option 197, 198, 202, 205-207
   grouped bar chart example 202
   LEGEND= option 188, 189, 191, 205
   MAXIS= option 171, 186, 188, 189, 200, 202
   MIDPOINTS= option 172, 188
   RAXIS= option 171, 186, 188, 189, 200, 202
   slash (/) separating variable name from options 171, 178, 187, 202
   specifying chart variable 171
   stacked bar charts, example 183-184, 187-188, 190
   SUBGROUP= option 183, 187, 189, 190, 191, 205-206
   SUMVAR= option 171, 174, 176, 188, 190, 202, 203
   TYPE= option 176
   vertical bar chart example 171

vertical bar charts, creating 167-179
   altering sequence of bars 172
   capacity utilization chart, example 172
   charts without DISCRETE option 175
   charts without SUMVAR= option 174
   combinations of SUMVAR= and TYPE= options 176
   controlling appearance of chart 177
   hourly utilization chart 168
   nongraphic graphics 173
   program for creating 169-171
   summary 178

W

$w. format 51
WEEKDATEw. format 87-88
   ordering of days 88
   output, examples 87, 88
   purpose and use 87
WEEKDAYw. format 87
WEIGHT statement, FREQ procedure
   calculating percentages 131-132, 134
   summing values 134
WHEN statement 106
WHERE command 41, 42
WHERE statement
   compared with subsetting IF statement 126-127
   logical operators 77
   purpose and use 77
   subsetting data for MEANS procedure 76-77
width of formats 51
WORDSw. format 51

Y

YEARw. format 87
YYMMDDw. format 51

Z

Zw.d format 51

Special Characters

$ (dollar sign)
   See dollar sign ($)
( ) (parentheses)
   See parentheses ( )
* (asterisk)
   See crossing operator (*)
, (comma)
   See comma operator (,)
/ (slash)
   See slash (/)
: (colon)
   See colon (:)
= (equal sign)
   See equal sign (=)
   See equal sign modifier (=)
@ (at) sign
   See at sign (@)